101 THINGS
YOU SHOULD KNOW ABOUT
2012

Countdown to Armageddon ...
or a Better World?

MARK HELEY, MA

AVON, MASSACHUSETTS

Published by
Adams Media, a division of F+W Media, Inc.
57 Littlefield Street, Avon, MA 02322. U.S.A.
www.adamsmedia.com

ISBN 10: 1-4405-1113-6
ISBN 13: 978-1-4405-1113-4
eISBN 10: 1-4405-1132-2
eISBN 13: 978-1-4405-1132-5

Printed in the United States of America.

10   9   8   7   6   5   4   3   2   1

Library of Congress Cataloging-in-Publication Data
Heley, Mark.
101 things you should know about 2012 / Mark Heley.
p. cm.
Includes bibliographical references.
ISBN 978-1-4405-1113-4
1. Two thousand twelve, A.D. 2. Maya calendar. 3. Maya astronomy. 4. Maya philosophy. 5. Earth—
Forecasting. 6. End of the world (Astronomy) I. Title. II. Title: One hundred one things you should
know about 2012. III. Title: One hundred and one things you should know about 2012.
F1435.3.C14H45 2011
909.83—dc22
2010039613

Contains material adapted and abridged from *The Everything® Guide to 2012* by Mark Heley,
copyright © 2009 by F+W Media, Inc., ISBN 10: 1-60550-161-1, ISBN 13: 978-1-60550-161-1.

This book is available at quantity discounts for bulk purchases.
For information, please call 1-800-289-0963.

Contents

INTRODUCTION
Have You Heard?

By this point, you've probably heard a thing or two about 2012. After all, it was the topic of a big-budget, star-studded movie that came out in 2009. After watching the movie, reading a magazine article, or talking to friends, you might have heard a variety of theories about what the year 2012 will bring—some of which may cause you to shake in your boots or shake your head in disbelief. You might wonder what the ancient Mayan people could possibly have known about science way back when, and how their calendar could have any bearing on today's world. If so, you've come to the right place. This book includes all the information you need to understand what this 2012 business is all about.

The basic fact: December 21, 2012, is the last date on the Mayan calendar. Why does this matter? Well, the Maya were very sophisticated astronomers and mathematicians. Their calculations of the cycles of eclipses and the motions of Venus are almost as accurate as those made by modern astronomers, even though they were made with the naked eye many hundreds of years ago.

Many theorists believe that the end of the Mayan calendar—December 21, 2012—marks the end of an old cycle and the beginning of a new world age. The Maya certainly believed that the world had been destroyed several times previously, and in each case the world was remade so that life could once again continue. Depending on who you talk to or what you read, this could mean many things. Some predictions are moderate, including varying degrees of climate change, something we're already experiencing on Earth. Others suggest events on a larger scale, such as galactic superwaves, which are massive ripples in gravity that are powerful enough to ignite supernovae in their paths. Events of that magnitude would have catastrophic effects on humankind, and could even cause the end of Earth as we know it. Don't panic, though; these sorts of events are highly unlikely. While they are possible and even probable in our planet's future, chances are they will not happen in the *near* future.

It's no secret that our world is currently experiencing changes that will have a significant impact on the way we live. From the strain on natural resources due to booming population growth to the extinction of certain species due to climate change, it's clear that we've reached a critical moment in our history. There's no better time for us to review humanity's relationship to the planet, and in essence, that is what this book is all about. In addition to learning about the various scientific theories and predictions, you'll read about ways that you can start making a difference in your own community. Such efforts can have a ripple effect, positively influencing other communities and, eventually, the whole world.

Can one person or one community really prevent catastrophic planetary events? There's only one way to find out. . . .

PART ONE

Who Are the Maya?

The Maya are a group of tribal people who live in an area of Central America stretching from the Chiapas region of Mexico south through Guatemala and modern-day Honduras. The indigenous Maya still follow in the footsteps of their ancestors, and though they have been subjected to severe persecution over the centuries, they have kept their traditions and culture alive.

Learning about the Maya is the first step to unlocking the mystery of 2012.

1. When Did Mayan Civilization Thrive?

The Mayan civilization flourished for about 600 years, from around A.D. 300 to A.D. 900. This time is usually called the classic period. The beginning date of the classic period roughly corresponds to the first time that an inscription of the Mayan calendar appears. The ending date corresponds to the last point at which distinctively Mayan architecture was being built.

During this period, the Maya built a series of city-states typified by elaborate complexes with step pyramids, ball courts, and other buildings sometimes called monasteries or palaces. The centers of the cities consisted of large ceremonial structures. The pyramids and other buildings were ornately decorated with features such as stylistic heads of jaguars or rain gods; murals featuring scenes of battle, the taking of prisoners, and human sacrifice; date inscriptions in hieroglyphs; and statues of kings and gods.

The city centers were inhabited by the upper class and religious elite, who were responsible for conducting the ceremonies that the rest of the Maya relied upon. The majority of the Maya lived in small buildings that clustered around these centers, and worked in fields and gardens that radiated out into the rainforests and mountainsides. They would come into the city centers to hear the oracles spoken by the priests and to participate in the frequent ceremonies that made up Mayan life. They believed these practices were necessary to appease their gods and ensure that the harvests would be fruitful and plagues and famines would be avoided.

This aristocratic religious class practiced bloodletting, astrological divination, and sometimes human sacrifice on behalf of the gods. They wore elaborate headdresses and ornamental costumes, made pottery, and carved statues in stucco and stone. In order to cleanse themselves for a ceremony, they would offer a sacrifice involving blood every day for a specified number of days leading up to an event. Techniques included piercing the ears, passing a cord of thorns through the tongue, and even drawing blood from the penis using a stingray spine!

2. Time Was the Center of Mayan Life

For the classic-era Mayan civilization, the measurement of different cycles of time was an all-consuming passion at the very center of their culture. In fact, the development of the Mayan form of writing, a system using symbolic pictures called hieroglyphs, was based on a desire to record different moments in time as accurately as possible.

The way the Maya viewed time is radically different from the way we in the industrialized world see it today. For the Maya, each day was not just a moment in time, but also a period watched over by a particular god. The god of the day would carry that time as his burden before passing it on to the next. Each god had particular influences, making one day good for hunting, for example, and each day's qualities were a unique combination of the influences of the presiding divine energies. Mayan priests and prophets would try to determine the influence of each god in order to choose the best course of action.

The Maya believed that if you had enough information about each time cycle, you could use it to see into the future. Certainly, the practice of prophecy was associated with rites and rituals, but at its core was observation of the cycles of nature. In essence, Mayan prophecy was as much a science based on the world around them as it was a religious practice.

Many of the surviving Mayan prophecies are depressing. They warn of bad harvests, disease, and political turmoil. In the midst of these, there are a few more favorable ones—such as spiritual awakenings—but they are the minority. Then again, if you took all of today's newspaper headlines of one year and used them to predict the events of the next year, it would probably make for pretty grim reading, too. This is the point of prophecy. If life were inherently stable, there would be little of value to report. Good news is not very often big news. It was the big events that the Maya were interested in predicting, and these, by nature, were more likely to include conflict and disaster.

3. The Mayan View of Time Differs from Our Own

To better understand the Mayan view of time, it's helpful to examine how the most popular current calendar came about. To do this, let's take a look at the European view of time. The life of a seventh-century European during the height of the classic-era Mayan civilization would have been dominated by saints' days and the seasons, as well as the traditional pagan holidays of equinoxes and solstices. The past and the future were still fuzzy concepts. At best, they were ideas discussed among scholars arguing about the date of the world's creation and its inevitable ending.

Most people measured periods longer than a year by the length of the reign of the current king. Because they believed that the earth was flat and that the sun and moon revolved around it, their view of the world and its sense of time were human-centered.

At the height of the classic-era Mayan civilization, on the very northern edge of Europe in a monastery, a scholar named the Venerable Bede was reinventing the Roman idea of *anno Domini*. First proposed by the Greek scholar Dionysius Exiguus, this dating system creates a linear chronology. It places Year 1 at the year of the birth of Jesus and dates everything in reference to that. Bede popularized this through his book, *Ecclesiastical*

Honoring Time the Mayan Way

Huge amounts of the Maya's resources were focused on keeping the necessary observations of time. The recording of dates on carved monuments, called *stelae*, spread like wildfire around the different Mayan centers, as each one tried to come up with the grandest, most elaborate recordings of the exact time. Auspicious or lucky dates were so important that the birthdates of rulers were occasionally changed to fit the best possible cycles. Accession to a throne was determined by the best date. This belief in the power of timing may even have been the original reason for the offering of human sacrifice, which may have started with the self-sacrifice of kings attempting to fulfill their divine roles by dying on the "best" day possible.

History of the English People, which was completed in 731. This convention not only established the "timeline" against which all dates are now measured, but it created a defined sense of a past and future. This established time as having a "before" and "after," rather than as just a repeating cycle, like the seasons of the year.

Though they seem like laws of nature, our laws of time are essentially theological decrees. They shape a remarkable amount of our experience, but, nonetheless, they are just conventions.

4. The Maya Disappeared circa A.D. 900

It's hard to imagine that such a rich culture could have so limited a life span, but by around A.D. 900, the Mayan civilization was in steep decline. New building stopped and the pyramids and ball courts were gradually abandoned to the jungle. Around this time the lowland population dropped by around 90 percent. There has been a lot of debate about what caused the collapse of the classic-era Mayan culture. Some suggest the burdens of ritual warfare between city-states became too much, or that a great epidemic wiped out the population.

The most likely explanation is that the Mayan system of agriculture, which relied upon a method of clearing rainforest and burning the vegetation to enrich the soil, broke down under the weight of their population. As soon as the population got past a critical mass, the fields couldn't be left alone long enough for them to become fertile again. This led to a downward spiral: fields were overworked, and the carrying capacity of the Mayan system of growing corn simply collapsed.

The classic-era Maya's demise is a stark reminder that civilizations that use vital resources at an unsustainable rate face the same fundamental issues. However great the architecture, learning, or history of a people, if there is no food, there can be no civilization.

5. The Maya Counted by Twenties

The Mayan calendar is at the heart of many of the 2012 prophecies, and the Mayan system of numbers is at the heart of their calendar. To understand the calendar's beauty and power, it's helpful to compare the Mayan system of mathematics with our decimal system.

The first striking thing about the Mayan counting system is the fact that the Maya counted in twenties, rather than tens. This is called base twenty, or the vigesimal system. This number base differs from the Arabic numbers of the decimal system that we're used to, which uses the base ten system.

In the decimal system, the more zeros that follow a number, the more times it is multiplied by a factor of ten. For example, the base ten or decimal number 786 can also be written like this, showing the ones, tens, and hundreds places:

$$(6 \times 1) + (8 \times 10) + (7 \times 100) =$$
$$6 + 80 + 700 = 786$$

Here's another example. The decimal number 3,440 can be written:

$$(0 \times 1) + (4 \times 10) + (4 \times 100) + (3 \times 1,000) =$$
$$0 + 40 + 400 + 3,000 = 3,440$$

Each extra column adds a factor of ten: increasing from ones to tens, then hundreds, thousands, tens of thousands, hundreds of thousands, and millions.

In the vigesimal number system, each position in a number raises the number by a factor of twenty, rather than ten. The Maya used columns of glyphs to represent the position of their numbers, rather than numerals. The higher up the column the glyph is placed, the higher the multiple of the number. Representing these vigesimal numbers in Arabic numerals looks something like this:

In base twenty, the vigesimal number 786 looks like this:

$$(6 \times 1) + (8 \times 20) + (7 \times 400) =$$
$$6 + 160 + 2,800 = 2,966 \text{ in base ten}$$

In base twenty, if we were to write the number 3,440, it would look like this:

$$(0 \times 1) + (4 \times 20) + (4 \times 400) + (3 \times 8,000) =$$
$$0 + 80 + 1,600 + 24,000 = 25,680 \text{ in base ten}$$

Instead of ones, tens, hundreds, thousands, and tens of thousands, the Mayan number system went from ones to twenties, four hundreds, eight thousands, and one hundred and sixty thousands. By the time the fifth position of a number is reached, the factor of that number is sixteen times bigger than it would be in the base ten. Numbers get much bigger much more quickly than in the decimal system. Thus, it gave the Maya the ability to conceptualize very big numbers and to add to and subtract from them very quickly.

6. The Maya Had a Concept of Zero

The Maya had both a concept and a symbol for zero, something the Romans never had. Zero didn't exist in the West until the Arabic numbering system was adopted around A.D. 1000. The Mayan zero is represented as a shell, meaning that the position in the number is empty, just like the number 0 in Arabic numerals.

The Mayan symbol for one is a simple dot. For the number two, just add another dot. Add more dots for three and four. The number five is represented as a bar. Six is simply another dot placed above the bar, and further dots are added in the same way for seven, eight, and nine. Ten consists of two bars. Eleven adds another dot above those two bars and so on, all the way up to nineteen, which is written as three bars with four dots above them. Nineteen, like nine in Arabic numbers, is the highest number that can occupy any position. Twenty is represented by introducing another position.

2012 PERSONALITY:
Malcolm Gladwell

In his book *Outliers*, social psychologist and author Malcolm Gladwell suggests that one of the factors that make Asian children good at math is that the words for numbers in Asian languages are generally shorter and more logical than those in Western languages. This seemingly small advantage translates into a huge effect over time. What Gladwell shows is that these tiny differences in the speed of computation add up cumulatively to a significant overall advantage. Likewise, the Mayan vigesimal system may have had even greater advantages because of its inherent simplicity and logic.

In the Maya's case, it was represented by adding another column of glyphs.

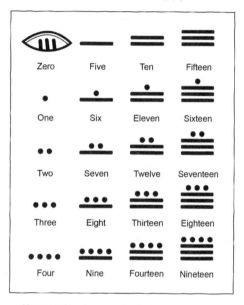

▲ Mayan numbers from zero to nineteen

This is a very elegant number system. Having just three symbols makes it one of the simplest ever devised. There is the shell symbol for zero, the dot represents one, and the bar stands for five. It's capable of representing big numbers very easily, and it doesn't require the memorization that the abstract symbols of the Arabic numerals from one to nine do.

7. The Importance of the Twenty Day Signs and the Tzolkin

For most Westerners, the week, the month, and the year are the most familiar and important divisions of time. For the Maya, however, a repeating cycle of twenty days was paramount. Each of the twenty days corresponds to one of the numbers from zero to nineteen in the Mayan vigesimal counting system. Each of the days also has a glyph or symbol that corresponds to it.

Tzolkin

The Yucatec Mayan word for the 260-day count.

Each one of the day signs marks not only a quantity of time but also a quality of time. Each also has its own specific energies and connections to plants and animals. Some were considered lucky, others dangerous. They were seen not just as units of measurement, but as living entities or gods. Each of these twenty day signs has thirteen different variations, which are represented by an added number. These variations repeat to make the first bigger cycle of the calendar. Each of the numbers also has its own qualities, so the combination of the numbers and the day signs create 260 distinct possible energies.

This 260-day cycle is now called the Tzolkin, a combination of the Yucatec Mayan words *tzol*, meaning "to count," and *kin*, meaning "a day." This word was made up by modern academics; no one

▲ The twenty day signs of the Mayan Tzolkin

knows what the classic-era Maya actually called it. The Quiché Maya, among whom the traditions of keeping the calendar are still very strong, call it the *Chol'qij*. (For simplicity, I will call the 260-day cycle the Tzolkin, except when referring to the traditional Quiché practice of day keeping.) The Tzolkin is the sacred almanac from which predictions and prophecies can be made. The twenty day signs and their names in Yucatec Mayan are on the facing page.

See #19 for more information on these days.

8. Mayan Numbers Have a Biological Connection

The significance of the numbers of the Tzolkin cycle is directly related to our physical bodies. The 260-day cycle is roughly equivalent to the cycle of human pregnancy. The twenty day signs also match our twenty extremities: ten fingers and ten toes. The thirteen numbers correspond to the thirteen major

When Does the Tzolkin Begin?

Historians disagree about the day on which the Tzolkin calendar begins and ends. One view is that, because it is a repeating cycle, it is possible to start and end anywhere. The most common convention, however, is to start on the day One Imix or Alligator and to finish on the day Thirteen Ahau or Sun. The traditional Quiché day keepers of Guatemala begin their count on Eight Batz or Thread. This is the day corresponding to Eight Chuen or Monkey in the Yucatec. On this day the Mayan day keepers hold one of their most important ceremonies, the Wajshikib Batz, to mark the re-creation of the sacred 260-day cycle and to initiate new day keepers. Other tribes use different days.

Other Ways to Break Down the Tzolkin

Within the Tzolkin are significant smaller cycles. Every set of numbers from one to thirteen is called a trecena and is governed by the glyph that starts the series. The energy of that glyph is taken into account when doing astrological readings or divination relating to any particular day. Traditional day keepers like to hold ceremonies on the middle three days of these thirteen-day periods when the energies appropriate to this work are said to be most balanced. The energies at the beginning of the trecena are considered too weak and those at the end too strong.

The Tzolkin can also be divided by four, with each quarter corresponding to a direction and called a season. Or it can be divided by five, with each fifth sometimes being attributed to one of each of the five Mayan worlds of creation.

articulations of the body: our ankles, knees, hips, wrists, elbows, shoulders, and neck.

The harmonic of Tzolkin's 260-day cycle is also found elsewhere in the cycles of nature. The growth period of corn from planting to harvest is around 260 days. Also, at the latitude of the Maya, 260 days is the period between the zeniths of the sun (the highest point reached by the sun during its orbit around a certain point of observation), a very significant event for the Maya. The period for which Venus is visible (261 days) is also very close to one Tzolkin, and this fact was used in tracking the cycles of this planet. The synodic period of Mars, which is the amount of time the planet takes to return to the same place in the sky, is equivalent to just under three Tzolkin cycles. The Mayan records of these cycles of time are so accurate that study of Mayan calendar inscriptions has led to the discovery of astronomical cycles unknown to modern science.

Clearly, the Tzolkin was much more than just a way of count-ing; it was part of the fabric of Mayan identity. Until the time of the European invasion, the classic-era Maya took the glyphs of their days of birth as their given names. Though this practice has died out, there is still a traditional ceremony that happens when a child is 260 days old to mark the completion of its first Tzolkin round. The day sign of your birth is believed to be responsible for shaping the whole of your life path, and much divination and astrological reading is still done today to determine what days are most lucky based on the sign of your birth. In traditional Mayan communi-ties, it is common to consult a day keeper in romantic matters to make sure that the signs of the lovers are compatible. Among its many facets, the Tzolkin can be seen as a combination of a biologi-cal counting system, an astrological and divinatory oracle, and one of the most elegant astronomical calendars ever devised.

9. The Haab Year Equals the Gregorian Year

Just as Westerners break up the year into twelve months, the Maya also divide the year—but instead they divide it into eighteen peri-ods of twenty days each. The year is knows as the Haab, and each of the twenty-day sections is called a uinal and has a special glyph and name. The days of the uinal are numbered from zero to nine-teen. For example, the uinal of Pop starts on the day 0 Pop, also known as the seating of Pop, and finishes on the day 19 Pop. This gives a total of 360 days, which is a round divisible number—but somewhat short of the true length of the year, which is 365.2422 days. The Maya compensated for this by adding five extra days at the end of the year called the Uayeb. These were considered unlucky days, when the gods rested and the normal barriers between the underworld and the waking world disappeared. The Maya marked this five-day period with rites of purification, fasting, and prayer, and waited expectantly for the start of the next year.

Haab

The yearly 365-day calendar.

The Maya did not make any provision for the remaining quarter day of the year and they recorded no leap year. As a consequence of this, the beginning of the year drifted through the conventional year over time. At the time of the European invasion, the beginning of the Haab, on the day 0 Pop, was equivalent to the date July 26 in the Julian calendar. Over the passing centuries, the beginning of the Haab has drifted to the end of February.

This kind of arrangement would be unthinkable to most of us. Certainly it made the Haab ineffective as an agricultural calendar. For a while, archaeologists thought that the Maya just weren't able to work out the length of the year. This isn't the case, though, as some inscriptions actually record corrections for this backward drift. The reality is that it wasn't a big concern. Instead, they were interested in making sure that a bigger cycle called the calendar round tied up accurately. By leaving the inserted day of leap year out, they were able to achieve this.

10. The Calendar Round Merges the Tzolkin and the Haab

Believe it or not, the Maya also created yet another calendar. They meshed together the 260-day Tzolkin cycle and the 365-day Haab cycle to create a bigger calendar round cycle. The common factor between the two calendars is 18,980 days, which equals exactly seventy-three Tzolkins and fifty-two Haabs. Each day in a calendar round has a unique combination of the Tzolkin and Haab signs. Only after 18,980 days will the same Tzolkin and Haab combination be repeated. Think of this cycle, of very close to fifty-two years, as the Mayan psychological equivalent of our century.

The calendar round was longer than the average life span and was considered to be sufficient for day and date calculation for most civil purposes. If a person knew both the Tzolkin and Haab dates for a day, he would be pretty sure of the day's place in the calendar round. The synchronization of these two calendars was of extreme significance to the Maya,

2012 PERSONALITY: Other Mesoamerican Cultures

The classic era was a rich period for culture in Central America. The Maya were surrounded by several other strong civilizations, including the *Zapotec*, centered in Monte Alban in Oaxaca; the *Teotihuacán* near present-day Mexico City; the *Olmec* of southern Veracruz; and the *Totonac* of northern Veracruz. The twenty-day cycle of time is found not just in the Mayan calendar, but also in all the Mesoamerican cultures of antiquity. The earliest recorded inscription of these glyphs goes back to Zapotec sites San Jose Mogote and Monte Alban, which may date from between 800 B.C. and 400 B.C.

and much more important than having a fixed-year calendar. The reason for this is that many of the functions of the fixed-year calendar—such as accurately dating the equinoxes, the zenith passages of the sun, and eclipse cycles—were already being performed by following the 260-day cycle. In other words, they just didn't need it. Agricultural cycles were measured from these reference points and also in divisions of twenty-day periods from the first rain of the year, so the calendar round was sufficient for most purposes.

What the Maya liked about the calendar round was that everything came to a perfect moment of synchronization once every fifty-two years. This type of mathematical perfection was very appealing to the Maya. It was a predictable and useful way of measuring time for them and they celebrated the end of calendar round by holding big festivals. Entire new layers were often added to pyramids to prepare them for the next fifty-two-year cycle. They set great fires, and sometimes they burned all of the huts and wooden buildings in which they lived. It was a time of great anticipation as they waited to see if the gods would grant them another fifty-two years.

Additionally, a Mayan individual's fifty-second birthday was the point at which he was recognized as a true elder, having completed what was seen as an entire lifetime. He was then effectively reborn into a second life, though during the classic period, few would have been lucky enough to live until fifty-two.

11. The Long Count Is the Calendar at the Heart of 2012

When people refer to the Mayan calendar in reference to 2012, in most cases they mean a calendar called the Long Count, also known as the thirteen baktuns. This is a cycle of approximately 5,125 years that will end, by most accounts, on December 21, 2012. The beginning date of the Long Count corresponds to August 11, 3114 B.C. in our calendar. However, the first recorded inscriptions of the Long Count actually appear around 35 B.C. at the pre-Mayan sites of Chiapa de Corzo and Tres Zapotes. This was some 300 years before the Maya emerged as a full-fledged culture, so we know that the Long Count wasn't their invention. Even though the Maya inherited this knowledge, they took its application to spectacular new heights.

Over time, the Long Count calendar spread from city-state to city-state across the Mayan lands. Its influence eventually reached from Chiapas in the west to Guatemala in the east and the Yucatán peninsula in the north. The propagation and development of the Long Count was to become one of the greatest cultural achievements of the Maya. It became a defining characteristic of what it meant to be Mayan. City-states and their rulers competed to create the most ornate monuments and statues, incorporating significant and auspicious dates of accessions and conquests.

Baktun

A calendar period of just under 400 years.

Some of these intricately carved statues are huge, up to twenty-five feet in height, and contain many elaborate inscriptions on all four sides. The artistry of Mayan carving—at the sites Copan and

The End of the Mayan Civilization

Around A.D. 900, for reasons still debated, the Maya started to abandon their cities. The most popular explanation is that their slash-and-burn method of agriculture used for the cultivation of corn and other crops led to an ecological collapse. As a result, they could no longer feed their population. After the classic-era Mayan decline, the surviving Mayan cities were more open to attack and invasion. The coming of the Mexica and Itzá to the Yucatán peninsula marked the end of this pure form of Mayan civilization. Widespread usage of the Long Count calendar probably only survived for 100 years after this point. Under cultural pressure from their invaders, the Maya began to radically shorten the time periods they were interested in recording. Sometime before the European invasion, knowledge of the Long Count was completely lost.

Quirigua, for example—is surpassed in technique and scale only by the monumental works of the ancient Egyptians.

The Long Count became the unifying symbol of the Maya in the classic period, when they built their most spectacular pyramids and ceremonial centers. At this time, Mayan culture was equal to or surpassed the accomplishments of any other contemporary civilization, including those in Europe.

12. The Structure of the Long Count

The Mayan calendar, or the Long Count, is made up of five different units. The tun unit is 360 days long, a rough approximation of a Western year's length, but the Long Count doesn't have a New Year's Day, and it isn't tied to a particular starting day of the solar year. Each tun consists of eighteen uinals of twenty days each. This is exactly the same structure as the Haab calendar, where eighteen twenty-day uinals make up the main 360-day part of the year. Twenty of these 360-day tuns make a bigger unit called a katun that lasts 19.71 years.

The katun cycle was important for the Maya in making prophetic predictions, and each katun had its own qualities and characteristics. Twenty of these katuns make up one baktun. Each baktun was a period of close to 400 years—394.25 years, to be exact. The whole

Tun

A calendar period of just 360 days.

cycle consists of exactly thirteen baktuns in total. This can be further subdivided into 260 katuns. These, in turn, contain a total of 5,200 tuns, the equivalent of 5,125.37 years. The different units of the Long Count are:

- **Kin:** 1 day
- **Uinal:** 20 days
- **Tun:** 18 x 20 days = 360 days
- **Katun:** 20 x 360 days = 7,200 days
- **Baktun:** 20 x 7,200 days = 14,4000 days
- **Thirteen baktuns:** 13 x 14,4000 = 1,872,000 days

A particular day in the Long Count is written as a series of all of these units, starting with the biggest number first (baktuns) and working downward to the number of days that have passed in the uinal. The first day of the thirteen baktuns is written 0.0.0.0.0 in this form of notation. The last day is 13.0.0.0.0. It is widely agreed that the thirteen baktuns began on August 11, 3114 B.C. and that

10.5.0.0.0.

One of the last recorded dates of the classic period was carved on a stele (a commemorative stone slab) at the site of San Lorenzo. This was inscribed with the date of 10.5.0.0.0., or October 5, A.D. 928. That's ten baktuns, five katuns, and no tun, uinals, or kin—a date exactly 1,476,000 days after the beginning of the count. This was a particularly important day in the calendar, as it marked the ending of a twenty-year katun. At that time, the statue of the god of the old katun would have been replaced in the Mayan calendar temple with a statue of the new one, signifying the changeover from one period to another.

they will end on December 21, 2012. To make the dates complete, the Tzolkin and Haab dates are added to the end of the Long Count.

The beginning of the Long Count:

```
August 11, 3114 B.C. =
0.0.0.0.0 4 Ahau, 8 Cumku
```

The end of the Long Count:

```
December 21, A.D. 2012 =
13.0.0.0. 4 Ahau, 3 Kankin
```

One of the earliest known inscriptions on a dated monument of the classic Mayan period was found at Tikal. It records the date of July 6, A.D. 292, or 8.12.14.8.15 in the notation of the Long Count. That's equivalent to eight baktuns, twelve katuns, fourteen tuns, eight uinals, and fifteen kin from the start of the Long Count—1,243,615 days.

13. The Maya Measured the Cycle of Venus

The Maya recorded the cycles of several planets, but Venus was particularly important to them, as it was regarded as the celestial equivalent of the god Quetzalcoatl (see #22). A whole table of calculations regarding the cycles of Venus can be found in one of the three surviving Mayan books, the Codex Dresdensis (Dresden Codex).

The Maya discovered that it was possible to follow multiple cycles of time simultaneously by using the 260-day cycle as the common factor. With Venus, which has a 584-day synodic period (the amount of time the planet takes to return to the same place in the sky), they were able to calculate that 146 Tzolkin rounds would be exactly equivalent to 65 Venus revolutions. This meant that 37,960 days elapsed between incidences of Venus rising as the morning star. Venus rising as a morning star took place on 1 Ahau, its sacred Tzolkin day. This was also exactly 104 years of their 365-day Haab calendar, or two calendar rounds.

These calculations are remarkable and would have taken many generations of observation to establish. It was an incredible achievement for naked-eye astronomers, subject to the vagaries of weather, to record these events so accurately. Without the sophistication of the vigesimal calendar, especially in the form of the Long Count, none of this would have been possible. At the peak of their culture, the astronomy of the Maya was at least as developed as that of the Spanish invaders, and the Maya did not have the advantage of the invention of the telescope. It is thought that the Maya tracked

2012 PERSONALITY:
Linda Schele

In the Codex Peresianus (Paris Codex), one of the surviving Mayan books, researcher Linda Schele has identified a Mayan zodiac dividing the sky into thirteen signs, rather than twelve as used in Western astrology. Another example of these interesting mathematical breakdowns is found in cycles the Maya were unable to track because they are not visible to the naked eye. For every twenty Venus rounds, there are twelve conjunctions of Uranus and Neptune. This cycle also equals exactly 2,080 Haab years or forty calendar rounds. It also makes one zodiacal age in the Mayan thirteen sign zodiac.

the stars by watching their motions reflected in specially built pools in their ceremonial centers.

The fact that so many planetary cycles can be harmonized by using a combination of the Tzolkin and the Long Count strongly suggests that the Mayan vigesimal counting system is more in tune with the laws of nature than the decimal one. It has long been established in astronomy that the orbits of the planets closely correspond to whole-number harmonics. This principle is called Bode's Law. It can be used to predict where a planet's orbit is likely to fall and also what the mass of a planet is likely to be. The Mayan calendar seems to articulate a similar principle.

Within the Long Count inscriptions, the Maya recorded even more cycles, including:

- A special glyph that was used to record the phase of the moon.
- A nine-day cycle that corresponded to the nine lords of the night and

a seven-day cycle that probably corresponded to gods of the earth.

• An 819-day count that was added at Palenque around A.D. 670 that seems to relate to the cycles of Jupiter and Saturn.

The Long Count brought these different cycles together in one big picture that could be used as a "snapshot" of a significant calendrical moment. The Maya could then capture this forever in stone.

14. The Short Count and the *Books of Chilam Balam*

When the Itzá and Mexica influences became dominant in the Yucatán around A.D. 1000, the keeping of the Long Count, which had already started to fall out of usage, went into serious decline. This was largely because the Itzá only used the fifty-two year-based calendar round. In a compromise between the different cultures of the two calendars, it was the Mayan Long Count that lost out. What replaced it was the Short Count. This abbreviated version is often illustrated as a wheel of thirteen katuns that make a circle of 260 years. The Mayan priests just shifted their focus from the Long Count cycle of the thirteen baktuns to the smaller scale of the Short Count of thirteen katuns and continued their ceremonies and divination as usual.

Both the Short Count and Long Count cycles end on the sacred day of Ahau, which is critical to the prophetic tradition of the Maya. The records of these Ahau katun prophecies were recorded in the

Books of Chilam Balam. The Chilam Balam were the hierarchy of jaguar priests that made divination and prophecy. The *Books of Chilam Balam* has several different versions. Each is named after the town the manuscript was found in, such as Chumayel, Tizimin, and Mani. They are the only actual written records of Mayan prophecy that still exist.

The *Books of Chilam Balam* contain some glaring errors, and the influence of the Spanish is evident in some places, but they are, in large part, authentically Mayan. The prophecies range from the reasonably positive to the near apocalyptic but tend toward the subject matter of daily life, such as famines, war, harvests, and political stability.

Chilam Balam

The order of jaguar priests responsible for prophecy.

The Chilam Balam were responsible for making the particularly important prophecy that accompanied each twenty-year katun period. These predictions were based upon what had happened in similar cycles and the results of their ritual divination. Each katun would be named after the number of the Ahau day on which it ended. For example, we are currently living in the katun 4 Ahau, the very last and two hundred and sixtieth katun of the whole Long Count. Some writers have argued that we can extrapolate from these katun prophecies and apply them to our own times. The results are often unconvincing. The evidence that the Chilam Balam still knew much about the 5,125-year cycle is debatable, but there are at least a couple of instances that seem to reference 2012 and the times we are now living in.

15. The Chilam Balam Talk of an Incoming Religion

One of the most quoted Chilam Balam prophecies talks about the coming religion of Hunab Ku. Hunab Ku is often translated as meaning "One God" but can also be interpreted as "One Sun" or "One Ahau." One Ahau is also the sacred day of Venus. Yucatec Mayan elder Hunbatz Men translates Hunab Ku as meaning the "giver of movement and measure." This prophecy is predicted for the katun thirteen Ahau, which is the last in the Short Count cycle of 260 years. This prophecy was given for the period A.D. 1544 to A.D. 1564, but it seems to speak about the future in general.

> "At the conclusion of the katun 13 Ahau, the Itzá will see . . . the sign of Hunab Ku, the erect tree which will be shown so that the world will be enlightened . . . confusion will be finished when the bearer of the cross comes to us."—*The Book of Chilam Balam of Mani* (Craine/Reidorp)

Most believe this refers to the coming of Christianity. The prophecy then asks the Itzá to receive the "bearded ones" as the messengers of a new god whose "commandments . . . will be good and (whose) new truth will be substituted for the old one. The Itzá will accept and worship the one True God who comes from heaven." A parallel is drawn between the god of the Spanish and the Mayan creator god Itzamna, and it seems to be a plea to the Itzá to adopt the faith of the invaders.

The Chilam Balam Versus the Itzá

In general, the prophecies of the Chilam Balam are disdainful of both the Itzá and the Spanish, even though the later ones are frequently addressed to the Itzá. The Chilam Balam themselves were largely drawn from the Tutul clan of the Xiu tribe, which may explain the snobbery toward the less calendar-literate Itzá. Though the times they record are often gloomy, the prospects of the future appear even worse. As one Chilam Balam, Natzin Yabin Chan, puts it: "Itzá, hate your gods, forget them because they will be destroyed by the foreign god you are going to worship."

A different version from Chumayel, however, adds the following paragraph:

" . . . they twist their necks, they wink their eyes, they slaver at the mouth, at the rulers of the land, lord. Behold, when they come, there is no truth in the words of the foreigners to the land."—*The Book of Chilam Balam of Chumayel* (Roys)

So the Chilam Balam seem to say something like, "Accept the religion of the Spanish, even though they are completely untrustworthy." What is confusing is that the Itzá and Spanish are fairly interchangeable in the prophecies. The same insults that had been used to describe the Itzá in older prophecies are even used to describe the Spanish, including the phrase "two-day ruler," a term used to denote the inexperience and lack of wisdom demonstrated by both sets of invaders.

16. Did the Chilam Balam Predict the End of the World?

There is a prophecy in one of the books of Chilam Balam that some think may relate to the end of the thirteen-baktun count in 2012. It talks about "tying up the bundle" of thirteen katuns but gives the date for this happening as 4 Ahau. The thirteen katuns actually end on a 13 Ahau day; it is the thirteen baktun calendar that ends on 4 Ahau. This is the date equivalent to December 21, 2012. The prophecy effectively predicts an end to the world.

> "In the final days of misfortune, in the final days of tying up the bundle of the thirteen (baktuns) on 4 Ahau, then the end of the world shall come and the katun of our fathers will ascend on high."—*The Book of Chilam Balam of Tizimin* (Makemson)

This is as close to a written prophecy about 2012 as can be found in the books of Chilam Balam, or in any of the surviving books of the Maya. It begins in the Makemson translation:

> "Presently the Baktun thirteen shall come sailing . . . bringing the ornaments of which I have spoken from your ancestors. Then the god will come to visit his little ones. Perhaps 'after death' will be the subject of his discourse."

This does suggest that the Maya thought the world would be destroyed in 2012, but they also believed that when the world was destroyed, it was recreated anew. This could be interpreted as a worldwide disaster rather than an apocalypse.

The prophecy finishes:

> "These valleys of the earth shall come to an end. For those katuns there shall be no priests, and no one who believes in his government without having doubts."—*The Book of Chilam Balam of Tizimin*

Essentially, this says that after 2012, we will be entering into a new time, uncharted by the Maya, for which "there shall be no priests." But this is not always interpreted as an age without religion. This example of the Chilam Balam's art seems to be the one that has the most relevance to the coming end of the Long Count in 2012.

17. The Fall of Mayapan

The prophetic tradition of the Chilam Balam went into decline after the fall of the last great Mayan capital, Mayapan. Mayapan's rule over the Yucatán ended after an army, thought to be that of Aztec king Montezuma, laid siege to the capital. The invaders announced that the Maya were to abandon their calendar and that it was now the katun of the spider. The Maya were also forced to worship a serpent called a sucking snake, considered an evil spirit. This effectively ended the succession of katuns and threw the priests of the Chilam Balam into disorder.

The Tortuguero Prophecy

There is one more Mayan prophecy that relates to the calendar end date of December 21, 2012: the Tortuguero Prophecy. This was found in an inscription carved on a stele monument at the site of Tortuguero. Mayan scholars have mentioned the existence of this stele in the past, but it is only recently that Geoff Stray, the author of *Beyond 2012*, has brought its real significance to attention. The inscription on the monument is badly eroded, but the legible part reads:

> The thirteenth baktun will be finished
> Four Ahau, the third of K'ank'in
> [Unreadable] will occur
> [It will be] the descent of the nine support gods to the [unreadable]

This inscription clearly gives the date of the calendar's end point. The nine gods may refer to the nine lords of the night, who, along with the thirteen lords of the day, were responsible for making the world. It is possible that the descent of the nine refers to an event happening at the ending of one cycle of creation and the beginning of another.

The end to the keeping of the prophecies came soon after the European invasion, but this was largely foreseen in the prophecies themselves. In fact, the Chilam Balam's predictions of what was about to happen seem to be very accurate. The European invaders' religion of Christianity did overtake the ways of the old gods. The Maya were hit with fasting and hardship, famine, and plague. The Chilam Balam predicted the Maya would go to war with the Spanish and would end up wearing their clothes and hats and speaking their language.

The Maya managed to continue to make prophecy despite all these changes, even when the new religion of Christianity and its Gregorian calendar was imposed on them. The Maya were not strangers to invasion, and they adapted to the new system just as they had when the Itzá had come. The priests instead made prophecies for the Christian year, basing their divination on which day of the week that New Year's Day fell on.

The content of these prophecies is very similar to the kind of predictions that were previously made for the old Mayan calendar cycles. This was the same tactic that the Maya had used to deal with invaders before—appearing to integrate on the surface level but doing so while keeping their traditions alive. Eventually, though, as knowledge of the calendar was lost, the traditions and stories of the old gods became blended with the saints and rituals of Christianity.

18. The Maya Are Still Around Today

Most of what has been discussed so far with regard to the Maya has come from the reconstructions of archaeologists and Mayanists. The ongoing collection of this body of knowledge has involved various academic disciplines and the dedication of many gifted minds over many decades of painstaking work. Yet the indigenous Maya of today have a living tradition that is just as important to learn from.

The Maya are a rural people who live mainly in the villages and smallholdings of the Guatemalan mountains and rainforest. Their way of life has remained largely unchanged since the European invasion. They have adopted Catholic saints in place of many of the old gods, but the feast days and qualities of their deities are often the same. The Catholic Church tried to win over the indigenous people to their form of worship by assimilating rather than destroying the local deities, so a blending of the two cultures has occurred over the last 300 years.

The Maya are adaptive by nature. During previous invasions, such as that by the Itzá and Toltec peoples in the Yucatán around A.D. 1000, they made considerable efforts to integrate the new cultures into the fabric of their own. Similarly, the Maya have attempted to bring the calendar and culture of Christianity into the Mayan way of life. As a result, the old traditions remain underneath a slightly different surface appearance.

In the remote villages around Lake Atitlán in Guatemala, tribal life has remained the least changed and the ancient traditions remain strongest. There is a variety of opinion among the

2012 PERSONALITY:
Day Keepers

The Mayan elders who are responsible for the keeping of the calendar are called day keepers. They keep the count of days by lighting candles and burning incense at the shrines set up to the specific gods of the day. These shrines are usually a short walk away from the village. On the appropriate day, the day keepers make the pilgrimage to these places to make offerings of corn and to burn the sacred incense copal, which is made from the resin of the ceiba tree. By making these offerings, the day keepers not only honor the gods but also keep them alive through the practice of remembering them.

traditional Maya about the sharing of their cultural heritage. For some of the day keepers, the calendar is considered knowledge that should be kept for the Maya and the Maya alone. They also believe the keeping of the count of days is a practice appropriate for the traditional elders alone. Only the initiated are allowed to speak the names of the day gods. To speak the names of the days is to invoke the gods themselves, and this must be done with full reverence and never lightly or in jest.

19. The Ceremony of Honoring the Directions

The honoring of north, south, east, and west is not just a fundamental ceremony of Mayan spirituality; it is a fundamental ceremony throughout the Americas. Almost all of the indigenous nations and tribes have their version of this universal tradition. All ceremonies usually begin with an acknowledgment of the directions and an honoring of the ancestors.

The elder conducting the ceremony begins by facing each of the directions and speaking a salutation or prayer to the spirits that reside there. This is usually accompanied by offerings of incense.

Words of gratitude and acknowledgement are then spoken to the spirits of the ancestors and the grandmothers and grandfathers that have gone before. The purpose of the honoring of the directions is to ground the ceremonialists in a sacred place and to connect them to rest of the world. This practice opens the space for further ceremonial work and establishes the proceedings as sacred.

The association of a particular color with each of the directions is very important to the Maya. Red is the color of the east, white the color of the north, black the color of the west, and yellow the color of the south. Each direction has further associations with specific birds, animals, and plants. These colors and their corresponding directions also apply to the twenty day signs of the calendar. The first sign, Imix or Alligator, corresponds to the east and is therefore red. The second, Ik or Wind, belongs to the north and is therefore white. Akbal or Night, the third sign, is black, the color of the west. Kan or Seed is yellow and belongs to the south. The succession then repeats four more times for the rest of the glyphs.

The Maya see the directions, like the days, as gods. They are called the Bacabs, and in the classic-period temples there are many inscriptions of them holding up the world in an Atlas-like fashion in each of their respective quarters. The red, eastern Bacab is called Likin; the white, northern Bacab is called Xaman; the black, western Bacab is called Chik'in; and the southern, yellow Bacab is called

Nohol. There is also a sacred ceiba tree for each of the directions of a corresponding color. Knowing and respecting the different energies of the directions is fundamental to the Mayan cosmos, and everything in the Mayan world is governed by their influences.

20. Who Is Don Alejandro Cirilo?

Don Alejandro Cirilo is the current elected leader of the National Mayan Council of Elders of Guatemala and a thirteenth-generation Quiché priest. As such, he is a very important voice for the indigenous Maya. He has been practicing the fire ceremonies and day keeping of his people's traditions his entire life. Cirilo, or Wandering Wolf, has been acknowledged by gatherings of indigenous elders from many of the American nations as a healer and medicine elder of considerable prowess.

Cirilo has spoken of a time that will be coming very soon, for which we should be preparing ourselves. It will be characterized by major changes in the earth, probably earthquakes and floods, but most significantly by a darkening of the sun. This period will last for several days and nights and people need to prepare both practically and spiritually for this by staying indoors, having enough supplies of food and water, and praying and purifying in readiness for the beginning of a new time. When the sun returns, this will signify the beginning of the new world age.

Cirilo has warned against unnecessary fear mongering and prophecies of doom. His advice on the coming changes is "Don't be afraid." Dramatic though the changes may be, he says they will not be an apocalyptic destruction of all life. By having a positive outlook and a willingness to examine our lives, he believes we can directly influence the outcome of these changes.

Cirilo is now very actively involved in international conferences and gatherings where he speaks on behalf of the Quiché Maya about their perspective on the calendar and the current times. The National Mayan Council of Elders of Guatemala has appointed a group of twenty-five elders who have been given the responsibility of sharing knowledge about the calendar and interpreting its significance. This information is expected to be released to the world before 2012.

21. The Mayan Elder Hunbatz Men Reopened Ceremonial Sites

Hunbatz Men is a Mayan elder of Itzá descent who lives in Merida, the largest city

The Institute for Cultural Awareness

To bring the message of the Mayan elders and their ceremonies to a wider audience, Cirilo has been working with the Institute for Cultural Awareness (ICA). The ICA has held several vision councils in which indigenous elders of different nations from throughout the Americas have come together to work ceremonially and to share their knowledge. The ICA is a nonprofit organization honoring indigenous tradition and is dedicated to providing a safe and healthy environment for cultural exchange and healing. They can be contacted through their website for information about future events and ceremonies (*www.ica8.org*).

2012 PERSONALITY: Carlos and Geraldo Barrios

Carlos Barrios is a Guatemalan of Spanish origin living at Huehuetenango who, along with his brother Geraldo, has studied with the local Mam tribe of the Maya for twenty-five years. He has been accepted by them as a day keeper of the Eagle clan of the Mam and speaks and writes on their behalf. His principal message is that the Mayan elders do not believe the world will end on December 21, 2012; they believe it will be transformed. This date will herald the beginning of the fifth world of creation, which will be marked by a return of a fifth element: ether. This new element will foretell a new way of being for humanity and will be accompanied by significant changes in the material world.

in the Yucatán. The Maya of the Yucatán live in quite different conditions from the traditional villages of Guatemala and are much more assimilated into mainstream Mexican society. Men has played an important role in reopening the ancient ceremonial sites to the indigenous people.

Until the beginning of the 1990s, the Mexican government was very uneasy about ceremonies at sites like Chichén Itzá, Uxmal, and Palenque. Native ceremony was seen as a challenge to the government and to the rule of law. Where traditional forms of worship were allowed, it was strictly controlled and, at many ceremonial sites, it was totally prohibited. Men took the lead in bringing together elders and other interested participants to reactivate these centers. The Mexican government resisted the ceremonies at first and insisted that some of the ceremonies be performed while surrounded by armed troops. The ceremonialists bravely faced these obstacles to burn copal at the pyramids.

By the spring equinox of 1995, the tide was turning and hundreds of thousands of people had gathered in the shadow of the Pyramid of the Sun at Teotihuacán near Mexico City. At the spring equinox in 1997, the Mayan elders were joined by a group of Tibetan lamas and held a series of solar initiations to join together these two peoples and celebrate their strong histories of ceremony and struggle.

Little by little, the ancient sites have opened up. Today, the spring equinox is celebrated at Chichén Itzá with a major festival that attracts many tens of thousands of people and has large international sponsors. Indigenous ceremonialists are at least tolerated at most of the sites, and the awareness of Mayan spirituality and its relationship to these great ceremonial sites has increased enormously.

Hunbatz Men still regularly takes groups of pilgrims on tours of the ancient Mayan sites to perform ceremonies, meditations, and healings. Many of the most important sites in the Yucatán and beyond have now been activated in this manner. He is the author of the book *Secrets of Mayan Religion and Science*, which details the importance of the actual sounds used in the Itzá Mayan language and the esoteric meanings behind them. Many of the words can be reversed to reveal related but different meanings. The Mayan language can be seen, Men claims, as a form of mantra, like the ancient Vedic languages of India.

Men has also worked on reconstructing the calendar traditions of the Itzá. The traditions of calendar keeping in the Yucatán didn't survive the European invasion, and Men had to reconstruct this lost knowledge from the fragments that are still known. This enterprise was only partially successful, and eventually, at a meeting of Mayan elders, Men agreed to adopt the traditional or Quiché count, effectively creating a unified count of the indigenous Maya.

PART TWO

Modern Interpretations of the Mayan Calendar

Now that you're familiar with the classic-era Mayan culture and the origins of the Tzolkin and the Long Count, it's time to fast-forward to today. One of the major reasons why you're reading this book is probably the fact that December 21, 2012, is rapidly approaching. But even if your interest in 2012 is new, people all over the world have been studying and interpreting the Mayan calendar and prophesies for decades.

Read on to learn what modern-day writers and scholars have to say about the Maya and 2012.

22. The Legend of Quetzalcoatl

Quetzalcoatl is an Aztec word meaning "feathered serpent." The Quetzalcoatl was a combination of a prophet, high priest, and king. Mayan legends say Ce Acatl Quetzalcoatl was originally from Tula, just north of present-day Mexico City, and that he came to the Mayan lands of the Yucatán after being driven out by his rival, the god Tezcatlipoca. What we know is that a great leader called Cuculcan entered from the West at roughly the same time the Itzá arrived, between A.D. 967 and A.D. 987.

▲ Quetzalcoatl dressed as the Lord of the Wind, from the Codex Magliabecchiano

Quetzalcoatl conquered the Yucatecan city of Chichén Itzá and established the beginning of a new culture that was a fusion of Mayan and Mexican/Itzá. He was seen as a great reformer and is also sometimes given credit for the ending of the practice of human sacrifice. At the end of his life, he is said to have floated out to sea on a raft made of serpents. His legend says he will return in a time of need. This is reminiscent of the stories and legends of King Arthur and should perhaps be taken as a blend of myth and oral history.

You may be wondering if Quetzalcoatl was a person, a god, or some kind of imaginary beast. After all, with a name meaning "the feathered serpent," it can be hard to tell. The answer, unfortunately, is not a straightforward one. Quetzalcoatl was a mythic figure but the name was also the title of the chief high priest and the given name of at least two known historical figures. Sometimes these figures are blended, creating a lot of confusion. To make things even more

2012 PERSONALITY: Tony Shearer

Tony Shearer, an author of Lakota heritage who was studying the culture and myths of ancient Mexico, first published his ideas in 1975 in his book *Beneath the Moon and Under the Sun*. Shearer believed that the period between the arrival of the conquistador Cortez, reputedly on Easter Sunday, April 21, 1519, and August 16, 1987, corresponded to a prophecy attributed to Ce Acatl Quetzalcoatl. This prophecy is about a cycle of thirteen heavens and nine hells, each of which would last a complete calendar round of fifty-two years, making the whole period last 1,144 years. The cycle of nine is a very significant one in Mayan cosmology and corresponds to the Bolontiku or nine lords of the underworld. These are nine gods that govern the cycle of darkness.

complicated, some contemporary authors have also claimed to channel the spirit of Quetzalcoatl and speak on his behalf.

23. What Does the Harmonic Convergence Have to Do with 2012?

For many interpreters, August 16, 1987, or the harmonic convergence, marks the beginning of a rapid period of transformation as we head toward 2012. It was Shearer who first noted the harmonic convergence date, but he saw it as the end of the fifth world of the Aztec prophecy and the beginning of the sixth world. It was Jose Argüellés, the author of the best-selling book *The Mayan Factor: Path Beyond Technology*, who coined the term "harmonic convergence" and helped turn it into a globally networked, grassroots event.

Argüellés had been a founding director of the Whole Earth Festival in the 1970s and understood how to reach out to people and capture the popular imagination. He believed that the date of the harmonic convergence began the final 26-year countdown to the end of the Mayan Long Count in 2012. Once the initial message spread, others repeated it and joined in. People were encouraged to go to places they considered sacred and gather with others to usher in a new time of global harmony. Many gathered at locations like Stonehenge, the Golden Gate Bridge, Mount Shasta, Sedona, and Glastonbury, all considered power spots.

The celebration of harmonic convergence was a significant success, but not many of the participants knew very much about the origin of the prophecy they were acting out or the end date of the Mayan calendar. It was Shearer's interpretation of the prophecies of Quetzalcoatl, himself an invader of the Mayan lands, that gave rise to this date, rather than the Maya themselves.

One of Jose Argüellés's ideas is that the thirteen-baktun cycle of 5,125 years can be read as a kind of road map of history. Argüellés takes the thirteen baktuns and projects them onto a matrix of the 260-day Tzolkin cycle. Since the Tzolkin is made of thirteen different combinations of twenty glyphs, it divides easily into thirteen columns. When the calendar is read like this, it provides an interesting picture of history. Known history, starting with the Sumerian civilization, begins remarkably close to the calendar starting date of 3114 B.C. Also around this date, the first pyramids were built in Egypt, and Stonehenge saw its first wave of construction. At the exact center of the cycle is the birth of the great teacher

Is the Date of the Harmonic Convergence Accurate?

Shearer's calculations of the date of the harmonic convergence appear to have been discovered intuitively, rather than mathematically. It has been claimed that Cortez arrived on the day Ce Acatl or One Reed in the Aztec calendar—which is the day associated with Quetzalcoatl. This claim, however, is not substantiated. The year 1519 in the Aztec calendar was Ce Acatl, so the idea is broadly correct. Yet there is a 117-day discrepancy between nine calendar rounds and the number of days between Easter Sunday 1519 and the harmonic convergence of 1987. Shearer explains this error by saying that it corresponds to the period between Cortez landing and his confronting Montezuma, the ruler of the Aztec empire. Nonetheless, the date doesn't have the same scholarly foundation as the December 21, 2102 end date of the Mayan calendar.

Gautama Buddha, suggesting a meditative central midpoint. History starts to speed up after this point, with the peak of Mayan civilization occurring halfway through the second half of the thirteen baktuns.

24. The Connections Among the *I Ching*, DNA, Benjamin Franklin, and the Tzolkin

For some, the speed and pace of modern times serves to back up the idea of December 21, 2012, as an end date. In the last several hundred years, we have seen inventions ranging from the steam engine to the nuclear bomb. At the same time, population has increased and the power of our communications technology has increased exponentially.

For Argüellés, December 21, 2012 represents an opportunity to transcend material technology and return to a more spiritualized, balanced culture. He claims that by using the Tzolkin (the sacred almanac from which predictions and prophecies can be made) we can free our minds from their historical shackles and develop a form of natural telepathy. The use of a calendar in tune with the cycles of nature can help facilitate this. According to academic Mayanists, there is no real evidence to indicate that people could begin practicing natural telepathy, but it has been a very popular idea about what makes the Mayan calendar ending in 2012 so important.

In his book *Earth Ascending*, Argüellés draws further parallels between the structure of the Tzolkin and the structures of the *I Ching* and DNA. He notes that both the ancient Chinese oracle of the *I Ching* and the modern discovery of the structure of DNA are based upon sixty-four distinct elements.

Argüellés also uses the mathematical structures known as magic squares to show a possible mathematical connection between the sixty-four permutations of the *I Ching* and DNA and the 260 permutations of the Tzolkin. (A magic square is an 8 × 8 grid in which all of the numbers from one to sixty-four are laid out.) In two particular combinations, each of the rows and columns add up to a total of 260. Benjamin Franklin was responsible for identifying these two magic squares, which have an ancient origin and were highly regarded by mathematicians of antiquity as providing insight into how nature is organized.

52	61	4	13	20	29	36	45	260
14	3	62	51	46	35	30	19	260
53	60	5	12	21	28	37	44	260
11	6	59	54	43	38	27	22	260
55	58	7	10	23	26	39	42	260
9	8	57	56	41	40	25	24	260
50	63	2	15	18	31	34	47	260
16	1	64	49	48	33	32	17	260
260	260	260	260	260	260	260	260	

64	2	3	61	60	6	7	57	260
9	55	54	12	13	51	50	16	260
17	47	46	20	21	43	42	24	260
40	26	27	37	36	30	31	33	260
32	34	35	29	28	38	39	25	260
41	23	22	44	45	19	18	48	260
49	15	14	52	53	11	10	56	260
8	58	59	5	4	62	63	1	260
260	260	260	260	260	260	260	260	

▲ The magic squares of Benjamin Franklin

This mathematical relationship is the springboard for an imaginative leap: Argüellés suggests that the Tzolkin is more than just

How Is DNA Like the Tzolkin?

The structure of DNA has always existed; Francis Crick and James Watson's discovery revealed a fundamental structure of biological nature that was always there, though unknown to us. In a similar way, Argüellés suggests, the structure of the Tzolkin, the 260-day constant, is a fundamental pattern of nature, not a human invention or contrivance. These two principles have related mathematical properties. Argüellés suggests they function in a way that creates a kind of weaving together. In the case of DNA, this manifests in its double helix structure. In the case of the Tzolkin, it is manifested in a weave in time marked by a sequence of days called galactic activation portals.

a calendar and handy counting tool. He believes that the 260-day calendar is actually a template that exists as an organizing principle in nature in the same way DNA does.

25. Did the Maya Recognize Galactic Activation Portals?

This sequence of days within the Tzolkin is another discovery of Shearer's, first published in *Beneath the Moon and Under the Sun*. He highlights a series of fifty-two days in the 260-day cycle that form a distinctive pattern on the grid. Each of the days is paired with another three to create a weaving pattern that runs through the Tzolkin.

The pattern of these fifty-two days, or galactic activation portals, exhibits the property of symmetry in two planes, one vertical and one horizontal, both of which are centered on the middle column. This means that each of the portal days is connected to three others that are its reflections in these two planes,

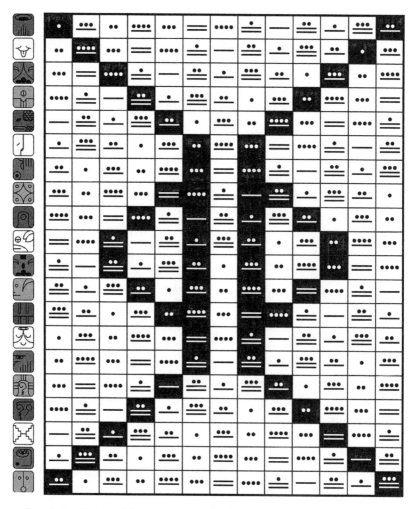

▲ The galactic activation portal sequence or Loom of the Maya

dividing the Tzolkin into four equal quarters. The significance of this is another interesting mathematical property, where the sum of each of the four linked portal days always adds up to the same number: twenty-eight. In this way, Argüellés shows that the Tzolkin functions like one of Ben Franklin's magic squares.

The researchers Geoff Stray and John Martineau recently uncovered another pattern in the Tzolkin that is based on the same harmonic numbers as the galactic activation portal sequence. It also uses groups of four days that are mirrored in the horizontal and vertical plane and add up to the sum of twenty-eight. However, it emphasizes the numbers one, seven, and thirteen in the Tzolkin. The result reveals a pattern that looks even more like the DNA double helix.

The magic of the mathematics that transfixed both Shearer and Argüellés demonstrates many interesting properties, but the evidence that any of these were actually used by the Maya themselves is pretty slim. Mayanists give no credence to the idea of galactic activation portals, and the idea that there were links between the Tzolkin calendar and *I Ching* is speculative. Argüellés's claims are far-reaching and visionary, but the elegance of the mathematical harmony that he demonstrates shouldn't confuse the fact that his ideas aren't provable.

26. Argüellés Creates the Concept of Dreamspell

In 1991, based on his work while writing *The Mayan Factor*, Jose Argüellés and his then wife, Lloydine Burrell, expanded the premise of the Tzolkin calendar as a harmonic matrix by creating a system of reading the signs of the Tzolkin. This became the toolkit *Dreamspell: Journey of Timeship Earth*.

Dreamspell is based on the concept that the Tzolkin, though a discovery of the Olmec and Maya, is actually a universal constant of time. It is, for the creators of *Dreamspell*, not the calendar's origins but the harmonic numbers found in the calendar itself that make it important and relevant today. Starting from this premise, *Dreamspell* rewrites the Tzolkin with the goal of creating an accessible and popular system. *Dreamspell* is intended to be usable by anyone, anywhere, living in any of the current cultures of this planet, to regain a connection to natural time.

For this reason, the day in *Dreamspell* follows the Gregorian start point of midnight. *Dreamspell* was designed to be a transitional calendar that allows an easy introduction to the energies of the Tzolkin. In *Dreamspell*, the definitions of the energies of each day are given in the form of just a few key words with universalized meanings. The idea behind this is that more people will become engaged in experiencing the calendar, rather than just reading about it.

With *Dreamspell*, Argüellés and Burrell take the interpretation of the Mayan calendar put forth in *The Mayan Factor* and rewrite it in a kind of mythological language. This makes their Tzolkin calendar into a fairy-tale-like game. Clear and simple interpretations of the meanings of each of the twenty Tzolkin glyphs are given, with just three keywords describing their essences. Each of the glyphs is also given an English name. These are based on the traditional

Argüellés's and Burrell's Philanthropy

One thing Argüellés and Burrell cannot be accused of is exploiting their interpretation of the Mayan calendar for money. The process by which they created or channeled the information for the *Dreamspell* led them to believe it should be given as a gift to humanity and should not be exchanged for money. Thousands of the boxed kits were given away for free, presumably paid for by philanthropic donors. This is because one of the core principles of their system is that we need to transition from "time is money" to "time is art." They believe that adopting their version of the Tzolkin would facilitate this. This stand became somewhat diluted later on as the *Dreamspell* gained in popularity and others wrote their interpretations of it, made calendars based upon it, or created T-shirts, pendants, or other time-is-art paraphernalia.

meanings, but in some cases Argüellés and Burrell universalize them. For instance, the traditional alligator or crocodile becomes the dragon. In *Dreamspell*, colors are attributed to each glyph according to their direction and correspond to those used traditionally.

In the early 1990s, Argüellés's writings were by far the most popular in introducing the concepts of Mayan calendrics to a wide public. *The Mayan Factor*, despite being densely written and full of unusual mathematics, made the *New York Times* bestseller list. When the *Dreamspell* box set was released, its combination of oracle and game found an audience eager to embrace it.

27. *Dreamspell's* Thirteen Tones of Creation

In *Dreamspell*, the thirteen numbers that are attributed to each day sign of the Tzolkin are reinvented as thirteen tones. This is reminiscent of the thirteen black and white notes in a musical octave; however, it must be noted that the Maya themselves didn't necessarily use tunings that fit to the Western scale. These tones or numbers are also given names that reflect Argüellés and Burrell's interpretation of their functions in this thirteen-note scale. This thirteen-note scale is called a wavespell, and is directly equivalent to a Mayan thirteen-day trecena. The following table shows the names *Dreamspell* gives to the numbers and day signs of the Tzolkin, along with the keywords that describe their functions and energies.

	Tone Name	Power	Action	Essence
1	Magnetic	Unify	Attract	Purpose
2	Lunar	Polarize	Stabilize	Challenge
3	Electric	Activate	Bond	Service
4	Self-existing	Define	Measure	Form
5	Overtone	Empower	Command	Radiance
6	Rhythmic	Organize	Balance	Equality
7	Resonant	Channel	Inspire	Attunement
8	Galactic	Harmonize	Model	Integrity
9	Solar	Pulse	Realize	Intention
10	Planetary	Perfect	Produce	Manifestation
11	Spectral	Dissolve	Release	Liberation
12	Crystal	Dedicate	Universalize	Cooperation
13	Cosmic	Endure	Transcend	Presence

Seal Name	Power	Action	Essence
Dragon	Nurtures	Birth	Being
Wind	Communicates	Breath	Spirit
Night	Dreams	Intuition	Abundance
Seed	Targets	Flowering	Awareness
Serpent	Survives	Life force	Sex
World-bridger	Equalizes	Death	Opportunity
Hand	Knows	Accomplishment	Healing
Star	Beautifies	Elegance	Art
Moon	Purifies	Water	Flow
Dog	Loves	Loyalty	Heart
Monkey	Plays	Magic	Illusion
Human	Influences	Free will	Wisdom
Skywalker	Explores	Space	Prophecy
Wizard	Enchants	Timelessness	Receptivity
Eagle	Creates	Vision	Mind
Warrior	Questions	Intelligence	Fearlessness
Earth	Evolves	Navigation	Synchronicity
Mirror	Reflects	Order	Endlessness
Storm	Catalyzes	Self-generation	Energy
Sun	Enlightens	Fire	Life

The combination of these three elements—the color of the direction, the tone, and the day sign or seal—make up the components of the *Dreamspell* name or galactic signature, for example, Red Cosmic Serpent. Similar to the ancient Mayan tradition of naming a person after the day on which he is born, this galactic signature is a person's key to playing the game of *Dreamspell*. The purpose of the "game" is not to win but to encourage a kind of cooperative play based on exploring the different harmonics of the calendar system. Doing this is supposed to allow the user to resynchronize with nature. Resynchronizing is a function of using the more natural timing system of the base twenty, vigesimal mathematics used by the Maya. Using the *Dreamspell*, Argüellés claims, leads to increases in the personal experience of synchronicity and, ultimately, to the liberation of human culture from the materialism of history.

28. The *Dreamspell* Controversy

Representatives of the Maya, including Cirilo and Barrios, have spoken out against the *Dreamspell*. The greatest point of contention is that the *Dreamspell* dates don't match up with the traditional Mayan calendar. This is very significant. Some people thought *Dreamspell* matched the Mayan calendar, and thus they became confused by the notion of two differing dates. In his defense, Argüellés does not represent his system as the Mayan calendar; but even so, it has become a popular misunderstanding.

Argüellés's Other Work

Jose Argüellés has gone on to create several other additions to his calendrical system. These include *Telektonon*, another board-game-like package that Argüellés claims is inspired by the great Lord Pacal of Palenque. Another box set called *7:7:7:7* attempts to integrate Russian plasma physics, Tibetan terma prophecies, and the Argüellés calendar, with some interesting results. He is currently living in Bali and is writing a hefty seven-volume series of books called the *Cosmic History Chronicles*, the last volume of which will be published in 2012. These books focus mostly on the more esoteric aspects of his work.

By using a different date, *Dreamspell* is out of line with the remarkable coincidence that the ancient classic Mayan calendar dates and the traditional indigenous Mayan dates match exactly. This means that not even a single day in this count has been lost for more than 1,000 years. This is a remarkable achievement, and it's difficult for the *Dreamspell* to justify having a different date. This has led many to suggest that it was simply invented. In an obscure set of essays called *The Rinri Project Newsletters*, Argüellés does address the issue of the start date and refers back to a date he uses as a correlation point. This correlation point was taken from a passage in the *Book of The Chilam Balam of Mani*, one of the important postconquest books of Mayan prophecy that you read about earlier. This date gives a specific Tzolkin date, a Haab date, and a Julian calendar date. This provides a correlation point to synchronize the calendars. It places the *Dreamspell* in the tradition of a continuation of or successor to the now-discontinued Yucatec Mayan tradition of keeping the Tzolkin calendar.

29. The Calendar Change Movement

After the end of the classic period and the conquest of the Mexica/Itzá, the Maya in the Yucatán went through a series of changes and innovations in their calendrical practices that led them to drop the Long Count in favor of recording multiples of the twenty-year katun cycle (the Short Count). They also experimented with changes that included stopping the count of the Tzolkin for thirteen days at the end of a fifty-two year calendar round to account for accumulated leap days. (The point of doing this was to make the actual solar year and Tzolkin match up perfectly, much in the same way that the 365-day Haab and Tzolkin previously had.) However, stopping the Tzolkin cycle for even a day is the greatest calendrical heresy imaginable for the traditional Maya of today. The unbroken count of days is one of the most important parts of their core spiritual practice. Yet, this was precisely what happened in the Yucatán after A.D. 1000. The Maya's desire to harmonize all possible time cycles, even those of their first conquerors, was the driving force behind the changes to the calendar. It is this ability to adapt and go with the flow that has helped the Mayan culture survive so much oppression. Although they lamented the Itzá's lack of calendrical prowess, they nonetheless tried to accommodate them. The most important calendar to the Itzá was the fifty-two-year calendar round. This system slowly overtook the 360-day tun-based counts, and eventually the knowledge of the Long Count was lost.

Today, Argüellés argues that the rediscovery of the importance of the Tzolkin should be the impetus to reform the world's calendar system. Argüellés and others involved in the current movement for calendar change see the ending of the thirteen-baktun cycle in 2012 as both a deadline and an opportunity. Their view is that our planetary crisis of overpopulation, environmental pollution, and global conflict is a byproduct of an inaccurate calendar. (The calendar we currently use is called the Gregorian calendar, named after Pope Gregory XIII. Because the calendar is used to determine the exact date of Easter, the Catholic Church steadfastly protected it as a sacred treasure. Hence, the calendar we use today was chosen through a display of religious and cultural power, not because it was the best or the most accurate.) While this might sound like a long shot, consider the fact that a calendar is a powerful cultural tool. Our view of time has important connections to how we behave. Those involved in the calendar change movement suggest that our current system treats time as an unlimited commodity, which results in a culture that takes the planet's resources for granted. Members of the movement feel that change is required at the root level of our culture, meaning that we must first change the calendar if we are to be successful in creating a more ecological culture.

Calendar change activists suggest that we adopt a "thirteen-moon" calendar that divides the solar year into thirteen moons of exactly twenty-eight days each for a total of 364 days; the extra day is taken as a "day out of time." (This day equates to July 25 in the Gregorian calendar.) What is useful about the number 364 is that two, four, seven, and thirteen all easily divide into it. This gives a year that can be halved and quartered and that has exactly fifty-two weeks every year.

As an added bonus, each month starts on the same day of the week. Another strange anomaly of the current system is that its months run independently from its weeks; a new month doesn't start on a particular day of the week. This makes it difficult to work out what day of the week a date will fall upon if it is more than a few weeks in advance.

30. Many South American Groups Favor Calendar Change

The idea of calendar change has become the focus of those using the Argüellés system of interpreting the Tzolkin. They suggest that the proposed calendar change is one way we can actually do

Counting the Cycles of the Moon

Despite its name, the thirteen-moon calendar is, in fact, a solar calendar. Why? There are two significant cycles of the moon: the sidereal, the time it takes to come back to the same phase, and the synodic, the time it takes to return to the same place in the sky. The sidereal period of the moon is roughly twenty-seven days and the synodic is roughly twenty-nine days. Neither of these corresponds to the moons in the thirteen-moon calendar. A moon in the thirteen-moon calendar won't start or end on a particular phase of the actual moon. The number of days in one of these moons totals twenty-eight. This is the whole number mean between these two moon cycles, so it is from this that the calendar gets its cycles. It is still possible to chart the progress of the Moon using this calendar, as each phase usually moves backward one or two days each moon.

something about the looming deadline of 2012. Adopting a new, more harmonic calendar means we can practically and psychologically prepare ourselves for a new time. This social movement has become particularly strong in South America. In countries like Brazil and Argentina, it has turned into a populist movement with hundreds of thousands of supporters.

Calendar-change advocates are also part of a broader movement toward ecological living and social change. This is seen in the creation of new ecocommunities and gatherings called vision councils. These events are organized by groups such as the Global Eco-village Network, the Rainbow Peace Caravan, and the thirteen-moon-based Planet Art Network. The goal of South American vision councils is to bring together indigenous elders, New Age speakers, activists, and others to share information. Everyone is encouraged to work together in a process of consensus decision making.

These gatherings are good examples of people living with awareness of 2012 as an organizing principle. The ending of the calendar is a major focal point and is accepted as a cultural norm. But instead of considering the end point of the Mayan calendar a dead end, these groups are using it as a stimulus for change that might otherwise not happen. The vision councils organize to create a network of new communities and new lifestyles that are more in tune with the planet.

31. Carl Calleman Interprets the Mayan Calendar Another Way

Carl Calleman, author of *The Mayan Calendar and the Transformation of Consciousness*, has developed another way of viewing the Mayan calendar. Calleman believes the cycles of time in the calendar are represented in physical form by the sacred architecture of the pyramids. There are certainly some correspondences. Many Mayan pyramids have either seven or nine levels. Calleman says that the seven-level pyramid is a representation of the sacred number thirteen that both the Tzolkin and the thirteen-baktun calendar are based upon.

If you treat these levels as a series of steps, there are a total of thirteen steps in the journey up the pyramid and then back down again. The qualities of each number can then be applied to the series of steps. Taking inspiration from the Aztec calendar, Calleman links each steps to one of the Aztec deities of the Thirteen Heavens. These are the gods traditionally associated with numbers one through thirteen by the Aztecs. He chose Aztec gods because the Mayan names and attributes for these deities are no longer known.

Each of these gods is seen as a ruling influence on each of the baktuns of the great cycle. The thirteen Aztec gods break down further into two groups. Seven of them have positive qualities and six have more negative ones. These alternate sequentially. The Calleman model divides these into seven days and six nights. The days of the Calleman model are auspicious. A similar structure to the wave of history can be found in the seven days and six nights of the creation story in Christianity, Islam, and Judaism. Calleman presents this as a discovery of how the underlying process of history can be described as a series of waves.

During the days, the development of consciousness pulses forward toward enlightenment. The nights are periods of repose and integration, often characterized by reversals of progress and darker times. Calleman uses this model to analyze history and to give a reading of humanity's progress and setbacks according to the different days and nights. This is a development of Argüellés's idea of the thirteen baktuns as a road map of history. What Calleman adds to this is the idea that it happens in seven pulses. These seven waves are interspersed with six necessary periods of consolidation, which he calls the wave of history.

32. Calleman Has a Unique View of the World Tree

Calleman has a unique take on what makes the Mayan calendar important. Unlike Argüellés, he doesn't think it is the special harmonics of the calendar system that are responsible for its uniqueness. According to him, the causal agent is not the calendar itself, but a growth process that derives from a "world tree" lying behind it. The pulses of the world tree are what move history forward in a way that is then coordinated by the different periods of the calendar. The two together create a holistic system of evolution. The idea of time emanating from the growth processes of a world tree is very different from the modern idea of time as a quantity to be spent or purchased. To better understand this idea, we have to go back to the mythological roots of the world tree.

Myths of the world tree exist in several different cultures. A notable one is Yggdrasil in Norse mythology. The Maya also have a strong tradition of the world tree. In fact, there is a sacred ceiba tree for each of the four directions that act as a world tree supporting the sky. There is probably one more tree for the center. This is the fifth direction that is represented by the color green. Though not visible in representations of the directions, Mayan myths talk extensively about this central world tree. The terrestrial world of the four directions that we

Meridian

A meridian is one of the vertical lines used to circumscribe the world, such as the prime meridian (Greenwich meridian).

exist in could be seen as a product or fruit of this original cosmic world tree that lies behind it as a primary source.

A good example of the Mayan world tree can be found on the famous inscription on the sarcophagus lid of Lord Pacal of Palenque. This relief depicts the great Lord Pacal reclining and looking upward toward a cross of four arms. It is considered to be one of the masterpieces of Mayan art and was discovered completely intact inside the Pyramid of Inscriptions at Palenque.

Calleman's idea is that the world tree actually corresponds to a meridian of longitude. This world tree is visualized as a hypothetical midpoint on the planet. From this point, he theorizes that the seven waves of history expand outward organically. So the closer geographically a culture is to the meridian of the world tree, the earlier it will be influenced by the wave of history.

The location of the meridian of the world tree is at the longitude of 12 degrees east. This line bisects Rome and Scandinavia and is noticeably distant from the Maya's world. Calleman thinks that this particular longitude best fits the observed phenomena of history. His idea of a world tree is very different from anything that has been recorded about Mayan belief and seems to have much more in common, at least in location, with the Norse world tree. Calleman has admitted he is influenced by the Norse idea and that, being Scandinavian, he "feels the pulse of the world tree more strongly than most."

33. Calleman's Nine-Level Pyramid of the Underworld

Calleman's second pyramid of time is a nine-level one that complements the first pyramid by describing an increasing acceleration of time that runs parallel to the wave of history. It is constructed by equating nine different-sized cycles of the calendar with each of the nine stories. This is more ambitious than the description of just the thirteen-baktun cycle contained in the previous pyramid.

This pyramid represents, according to Calleman, nothing less than a description of time from the beginning of the universe. The bottom level is a vast cycle of time equal to thirteen hablatuns. Each hablatun is 1.26 billion years, so this step is 16.4 billion years long. As the pyramid is climbed, both the length of the cycles and the size of the steps become progressively smaller. At the same time, each step also represents a stage in evolution, so that as evolution progresses, it happens in increasingly smaller periods of time.

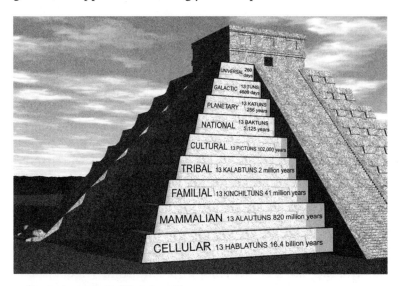

UNIVERSAL 260 days
GALACTIC 13 TUNS 4680 days
PLANETARY 13 KATUNS 256 years
NATIONAL 13 BAKTUNS 5,125 years
CULTURAL 13 PICTUNS 102,000 years
TRIBAL 13 KALABTUNS 2 million years
FAMILIAL 13 KINCHILTUNS 41 million years
MAMMALIAN 13 ALAUTUNS 820 million years
CELLULAR 13 HABLATUNS 16.4 billion years

▲ The nine-level pyramid of the Underworld

The closer we come to the end of the calendar, the quicker the shifts happen. For example, the bottom level establishes the material elements of the universe. The second level of 820 million years marks the beginning of life. The start of the thirteen baktun, on the sixth level, sees the establishment of written language. The seventh evolutionary leap, the era of industrialization, happens in just thirteen katuns, or 256 years. We have now reached the eighth level of the pyramid. This corresponds to the last thirteen tuns of the calendar. This is the period of just 12.8 years that we are now living in. The last cycle before we reach the top happens in the final 260 days and equates to the 260-day cycle of the Tzolkin.

In Mayan mythology, in addition to the gods of the Thirteen Heavens, there are also the nine gods of the Underworld. Each level of Calleman's pyramid corresponds to a Lord of the Underworld, so Calleman calls them underworlds. We are currently in the eighth or galactic underworld, in which Calleman sees the advent of telepathy and a transcending of the material framework of life. This will be followed by the brief universal underworld of 260 days that represents the evolution of cosmic consciousness and transcendence into timelessness. In this structure, all levels of the pyramid are added cumulatively until the highest level of creation is reached.

34. Did the Maya Know About the Big Bang Theory?

The inspiration for the pyramid of time comes from a Mayan inscription found at the site of Coba. The carving on this monument records a huge number of cycles, all of which are numbered thirteen. It goes beyond the thirteen-baktun count to include numbers hundreds of millions of times bigger than the known age of the universe.

Calleman's model uses the nine time cycles within this inscription that have been given names. The total time span of these cycles adds up to 16.4 billion years. This is approximately the current age of the universe, according to the big bang theory. (Estimates suggest this happened about 13.7 billion years ago.) This, Calleman suggests, means the Maya actually knew about the big bang and that the monument records this fact. This theory, unfortunately, ignores the fact that the names of the cycles are not actually from the Maya, but conventions that have been suggested by modern academic researchers.

It seems much more likely that the monument at Coba was intended to express the idea of a number so vast that it was beyond comprehension. The sheer size of the inscription suggests that the Maya recognized infinite cycles of time beyond counting. It's an enlightening insight into the Mayan philosophy of time but not really proof that the Mayan calendar refers explicitly to the big bang.

The appeal of Calleman's pyramid is that it shows a model of how time may be accelerating. This gives a

Calleman's Theory Does Not Mesh with Tuns

The tun is very inconvenient for Calleman's theory that the Mayan calendar is measuring something other than a physical amount of time. The tun is very much associated with a year and it is a fundamental unit of the calendar. In spite of this, Calleman claims that the Long Count does not measure astronomical time at all, but the spiritual evolution of humanity in its path toward enlightenment. He even goes as far as to say that it is not really a physical calendar at all. For him, the loss of the Long Count and its replacement by the Calendar Round in the postclassical period was responsible for a fall into materialistic consciousness. In his view, it is the Mayan calendar alone that has the true cosmic blueprint encoded in it.

framework for interpreting the trends of history in a way that may prove insightful. But there are further complications in the detail of this theory. For one, it ignores the fact that the cycles of the Mayan calendar are very much based around the 360-day unit of the tun. This is very significant because Mayan numbers usually jump by a factor of twenty. However, when we count the tuns in the Mayan calendar, this rule is broken, probably because the Maya wanted to capture the 360-day cycle of the sacred year in the count of days. At this point, it jumps eighteen—the number of the uinals of the year— rather than the usual twenty. This would make a pyramid with one different-sized step.

35. Calleman Disagrees with the End Date

One of the most notable aspects of Calleman's theories is that he disagrees with the established Mayan calendar end date of December 21, 2012. Despite his belief, working out what dates the Long Count inscriptions correspond to was the subject of extensive research throughout the twentieth century. It was a search that occupied the professional lives of dozens of researchers.

The process of finding the right date first involved looking at many different astronomical factors to find the best possible fit. Researcher Joseph Goodman suggested a correlation at the beginning of the twentieth century; Juan Martínez Hernández suggested another. In 1927, the great Mayan scholar J. Eric S. Thompson proposed a third possibility. In 1950, after further research, Thompson decided on one number, and at this point most academics agree with him.

His accepted correlation between the Mayan calendar and the Gregorian one has become known by the rather awkward name of the Goodman-Martinez-Thompson correlation. This is a combination of the names of the researchers who made the most significant contributions to its discovery. This is now widely adopted and fairly uncontroversial. It is also very significant that this correlation corresponds exactly to the date kept by the traditional indigenous Maya. There has never been any substantial debate about the correlation between the Tzolkin and the Long Count, which is what Calleman questions. There are hundreds of inscriptions and monuments that confirm the established dates.

Calleman has been very vocal in his criticism of *Dreamspell* because it changes the calendar used by the Maya. Ironically, his system does exactly the same thing by changing the end date of the Long Count. Altering this critical point means all the shifts that he forecasts are tied into an idiosyncratic timetable. Calleman's system ends on October 28, 2011, because it falls on the "auspicious" Tzolkin date of thirteen Ahau, a day that Calleman claims represents completion. This is out of sync with the accepted end date of December 21, which most academic experts on Mayan calendrics agree on.

Another reason Calleman picks thirteen Ahau as the end of the calendar is that it represents the energy of enlightenment. It is also the last date in the 260-day cycle in the most commonly used

Cirilo Backs Calleman's Ideas

Calleman's ideas do have one very significant backer: the Mayan elder Cirilo. Cirilo believes we're living within the time of twelve baktun, thirteen Ahau. This seems to fit with Calleman's idea of the calendar ending on thirteen Ahau. Calleman and Cirilo have also filmed an interview in which they discuss a prophecy of thirteen Ahau. The prophecy talked about in the interview refers to a katun prophecy and is taken from one of the *Books of Chilam Balam*. What Calleman and Cirilo have in common is that they are skeptical about the December 21, 2012 end date. Unfortunately, neither presents any evidence that brings the date of December 21 seriously into question.

convention. This is important to Calleman, as he believes the calendar is non-physical. Therefore, he does not regard the fact that the usual end date falls on a winter solstice in 2012 as being particularly significant.

36. Calleman Raises Awareness of Two Venus Transits

One big contribution Calleman has made to the debate about 2012 is his effort to raise awareness of two Venus transit events, in the summers of 2004 and 2012. These are very important events that would have been extremely meaningful for the Maya. Venus transits are associated with the return of Quetzalcoatl, and it is notable that the second one occurs in the year 2012 itself.

A Venus transit occurs when Venus passes across the face of the sun from the point of view of Earth. This creates an eclipse-like phenomenon where the disc of Venus can be viewed through a filtered telescope or blackened glass as it crosses the face of the Sun. Transits happen in pairs. Each of these pairs occurs approximately every 120 years. In the Mayan calendar, they are exactly eight Haabs of 365 days apart, which is very significant.

Calleman proposes that each of the pairs of Venus transit events represent breakthroughs in communication. This looks like a credible idea if we examine the last five sets of pairs:

- **1518 + 1526:** The period between the transits was marked by the circumnavigation of the globe by Magellan.
- **1631 + 1639:** The first national postal services began around this time (Denmark in 1624 and Sweden in 1636). This pair of transits is the first to be observed by telescope.
- **1761 + 1769:** On these two dates, the transit was the focus of the first truly global scientific experiment, as teams of scientists recorded the event simultaneously in many locations.
- **1874 + 1882:** These two dates mark the extensive use of the telegraph and the invention of the telephone.
- **2004 + 2012:** The year 2004 represent the global Internet revolution. Is 2012 the beginning of human telepathy?

Calleman suggests that a communication breakthrough will occur between the Venus transit pair of 2004 and 2012. He believes it will come in the form of the initiation of telepathy on the planet that will offer an opportunity to enable humanity to return to a state of oneness. It would seem that the continuing Internet

Celebrating Oneness

In order to promote appreciation of the transits, Calleman helped initiate an event called the Oneness celebration. The first of these happened on June 8, 2004, the date of the first Venus transit. The purpose of the Oneness celebration is to bring people together in meditation on oneness and the fundamental unity of all nature. A number of Oneness celebrations occurred around the world to celebrate the first transit. More events are planned for the second transit on June 5, 2012. There are also events planned for the calendar end date and the two solar eclipses that occur on May 20 and November 13 in 2012. These can be seen as successors to the original harmonic convergence event. They share the hope that by aligning events to these times, these gatherings will cause the shift in human consciousness necessary to avert imminent environmental catastrophe.

revolution would also be a good candidate for a significant communications breakthrough.

37. In the End, Calleman's and Argüellés's Theories Are Just That

The idea of the Mayan calendar as a timetable for evolution or map of history is interesting. However, the model put forward by Calleman raises many big questions. It attempts a comprehensive explanation, but it doesn't match the academic consensus on the Mayan calendar.

While his pyramid model may seem attractive as a possible model for showing the acceleration of time, it's really just an approximation. The usefulness of this model is also undermined by the revision of the end date. Furthermore, there isn't any substantial evidence to support the idea. It also makes the predicted shift dates out of sync with the accepted chronology of the Mayan calendar. In the eyes of many researchers, the weakness of the calendrical science behind Calleman's theory serves to further undermine the reputation of New Age versions of the Mayan calendar.

Interpretation and theorizing have been a big part of an attempted understanding of the workings of the Mayan calendar, first with Argüellés's *Dreamspell* system and then with Calleman's pyramid model. Neither are exact or accurate representations of the Mayan calendar, though both contain insight and interesting speculation.

This kind of speculation is probably necessary as part of a process leading to a greater appreciation and understanding of the significance of this extraordinary calendar system. The problem is that we in the West have a completely different view of time. Calleman's and Argüellés's ideas represent imaginative ways of attempting to see the world through the eyes of the Maya. In this way, they are probably helpful signposts or stepping stones along the route. However, neither of these models is a final understanding of the Mayan calendar.

What we might take from these insights is an inspiration to further examine the Mayan calendar for clues about how it functions. If these speculations could also be made to accurately reflect the actual cycles and dates that have been established by rigorous academic research, they would be on much stronger ground.

38. Why Bother Using the Tzolkin Calendar?

Ultimately, a calendar is a shared agreement about how to divide and organize time. By choosing to use a different standard of time, like the Mayan Tzolkin, parallel with the usual Gregorian calendar, it is possible to get interesting perspectives on both. Following the different energies of the day signs gives a different feeling to each day, and it also makes you aware of how much subliminal influence a calendar can have on our lives.

It does take a little work to gain an understanding of your place on the Mayan Tzolkin, however. One of the things that makes converting the date of the Gregorian calendar into the traditional Mayan Tzolkin more difficult is that there is no certain rule about what time the day begins and what time it ends. For some of the Maya, the day begins at dawn; for others, it begins at sunset. In neither case does it begin at midnight, as the Western calendar does. This makes a standardized approach to decoding your birthday's sign more complex because if you are born before dawn, or, in some traditions, after sunset, a different glyph applies.

Many thousands of people in dozens of countries around the world are now using the Tzolkin calendar in one form or another. Some have been introduced to the calendar in the traditional way, through meeting a day keeper and being given a reading. From the point of view of the indigenous Maya, this is still an essential experience. However, many more people have been introduced to the Tzolkin through books and websites about the calendar, events about it, or friends who already know about it.

39. Who Can Tell You Your Day Sign?

A reading from a traditional day keeper is a form of initiation or baptism, where a person is given her individual day sign, which also becomes her calendar name. Before the European invasion, all the Maya were known primarily by these names. This one-to-one transmission of knowledge goes back at least 1,000 years, but the Maya themselves believe their traditions date back as many as 5,000 years.

Not everyone is able to visit the Mayan highlands, where these traditions are still strong. Those who do, if they are lucky, may get the opportunity to learn directly from someone who has been born into the calendar. Few Westerners have been accepted into the traditions of Mayan day keeping. Not everyone is suited to such a role; those who are need patience and perseverance to learn the traditional way through a lengthy apprenticeship. The only way to become a day keeper is by living the calendar and immersing yourself in it and the worldview of the Maya.

One person who has been working in this way with the calendar and is now sharing his knowledge is Martin Prechtel, author of *Secrets of the Talking Jaguar.* Prechtel was raised on a Pueblo Indian reservation in New Mexico. His mother was a Canadian Native American and his father a Swiss paleontologist. He married a Mayan woman and raised two sons in the Mayan village of Santiago Atitlán. After a shamanic apprenticeship he became acting shaman of the community and eventually Nabey Mam, the first chief. The world he describes is a heart-centered one, rather than an intellectual one, where much of the wisdom contained in the calendar comes to a person not by study or contemplation, but by participating in ceremonies and the Mayan way of life. In the traditional communities, learning the calendar is as important as learning everyday folklore and the tales of the ancestors.

40. Finding Your Day Sign

Since finding a trained day keeper can be quite difficult, try figuring out your day sign on your own. There are two decoder charts to use to find a day sign. The first is for finding the *Dreamspell* interpretation of the Tzolkin. Use the second for finding a date in the traditional Mayan count. The method of reading the charts is the same.

1. Write down the birthday or date to be converted.
2. Add the year number to the month number.
3. Add to these two numbers the number of the day of the month that is being converted.

4. If the total of these three numbers is more than 260, deduct 260 from the total.

5. The result should be a number between one and 260. Look up that number on the Tzolkin chart. Each square of the Tzolkin has a number written in it. This is the number or tone of the date. Read across to the left-hand column. This symbol is the corresponding day sign or glyph.

6. Look up the meanings of the number and the day sign in the following section. The Tzolkin signature of the date is a combination of both.

Note that when using the traditional Mayan chart, if the year is a leap year, add 1 if the date is after February 29. If you know that a person was born after sunset on the day of their birth, use the following day sign in the traditional Mayan chart.

DREAMSPELL MONTH DECODER TABLE

Month You Were Born In	Number	Month You Were Born In	Number
January	0	July	181
February	31	August	212
March	59	September	243
April	90	October	13
May	120	November	44
June	151	December	74

DREAMSPELL YEAR DECODER TABLE

Birth Year	Birth Year	Number	Birth Year	Birth Year	Number
2013	1961	217	1987	1935	87
2012	1960	112	1986	1934	242
2011	1959	7	1985	1933	137
2010	1958	162	1984	1932	32
2009	1957	57	1983	1931	187
2008	1956	212	1982	1930	82
2007	1955	107	1981	1929	237
2006	1954	2	1980	1928	132
2005	1953	157	1979	1927	27
2004	1952	52	1978	1926	182
2003	1951	207	1977	1925	77
2002	1950	102	1976	1924	232
2001	1949	257	1975	1923	127
2000	1948	152	1974	1922	22
1999	1947	47	1973	1921	177
1998	1946	202	1972	1920	72
1997	1945	97	1971	1919	227
1996	1944	252	1970	1918	122
1995	1943	147	1969	1917	17
1994	1942	42	1968	1916	172
1993	1941	197	1967	1915	67
1992	1940	92	1966	1914	222
1991	1939	247	1965	1913	117
1990	1938	142	1964	1912	12
1989	1937	37	1963	1911	167
1988	1936	192	1962	1910	62

You can find a Traditional Mayan Year Decoder Table in *The Everything® Guide to 2012* or visit *www.mayanmajix.com*.

41. The Twenty Day Signs

Once you've found your day sign, you can learn more about the energy of that particular sign. On the following pages, you'll find:

- The names of the day signs in Yucatec Mayan
- The *Dreamspell* names
- The equivalent names in translation from Yucatec Mayan
- Three keywords from the *Dreamspell* describing the energy of the day

The first half of the interpretation of each day relates to the *Dreamspell* and the second half to the traditional Mayan interpretation.

How to Form a Day Name

The *Dreamspell* or traditional Mayan dates for a particular day can be put together to form a name. With the traditional Mayan calendar, this would include both the Yucatec Mayan form and its translation. For example, Kin 108 in the traditional count is Four Star or Can Lamat. The *Dreamspell* name is composed of the direction, followed by the tone, and then the day sign; so Kin 108 would be Yellow Self-Existing Star.

IMIX

Dreamspell: Dragon; **English translation:** Alligator
Keywords: Nurtures, Birth, Beginning
Color: Red
Direction: East

The Dragon begins the Tzolkin and represents new beginnings and fresh starts. The Maya consider the alligator or crocodile to be a primal being whose qualities represent existence itself. The sign of Imix, the alligator, is about birth and the forces connected to motherhood.

IK

Dreamspell **and English translation:** Wind
Keywords: Communicates, Breath, Spirit
Color: White
Direction: North

The Wind is symbolized by the tongue, the organ of spoken communication. It is also associated with the T-shaped holes that are found in many Mayan temple walls to allow the sacred wind to enter. The sign of Ik represents the breath of the wind as the carrier of divine communication. Ik is a sign of the air and of thought and ideas.

AKBAL

Dreamspell: Night; **English translation:** House
Keywords: Dreams, Intuition, Abundance
Color: Blue/Black
Direction: West

This sign is the house of the underworld, the domain of the Mayan Lords of the night and the dark place from which all dreams come. The Night is the imaginative realm where knowing is intuitive and still connected to the depths of the psyche. From this empty place, creation can spring forth, manifesting our dreams in the waking world.

KAN

Dreamspell: Seed; **English translation:** Lizard
Keywords: Targets, Flowering, Awareness
Color: Yellow
Direction: South

The sign of the Seed governs all the cycles of growth, from placing a seed into the ground through harvest. By targeting where we sow our seeds, we can ensure the best possible rewards. Those born on Kan, the day of the Lizard, are considered by the Maya to be dynamic, and the day sign has connotations of sexuality that might be expected from such a fertile sign.

CHICCHAN

***Dreamspell* and English translation:** Serpent
Keywords: Survives, Life force, Sex
Color: Red
   Direction: East

Serpent represents the kundalini power of life force. The base level of this energy is the survival instinct in all living things, but this can be cultivated to rise up the spinal column and become the force that drives our enlightenment. Serpent power is the reptilian mind in all its raw vitality. Chicchan people are said to have a strong sense of instinct.

CIMI

***Dreamspell*:** World Bridger; **English translation:** Death
Keywords: Equalizes, Death, Opportunity
Color: White
Direction: North

The World Bridger is the part that endures when a person journeys between life and death. It represents the power of transformation and the great opportunity that provides, if the choice is made to embrace it. The Cimi symbol of the skull of death is a great reminder that everyone is born equal and that, in death, all are made equal again.

MANIK

Dreamspell: Hand; **English translation:** Deer
Keywords: Knows, Accomplishment, Healing
Color: Blue/Black
Direction: West

The Hand is associated with the healing arts and the creative knowing that comes through touch. The Hand is representative of the wisdom of crafts and all that can be accomplished by the craftsperson's knowledge. The sign of Manik, the deer is considered by the Maya to be a powerful and spiritual day sign of high aspirations and ideals.

LAMAT

Dreamspell: Star; **English translation:** Rabbit
Keywords: Beautifies, Elegance, Art
Color: Yellow
Direction: South

The Star is associated with the planet Venus, a most important deity to the Maya. The Star is the sign of artists, responsible for beautifying and uplifting with their creativity. Those born under the sign of Lamat, the Rabbit, are considered by the Maya to be good gardeners and are likely to have a green thumb. The rabbit represents the struggle to overcome the material world and achieve spiritual liberation.

MULUC

Dreamspell **and English translation:** Moon
Keywords: Purifies, Water, Flow
Color: Red
Direction: East

The Moon is the sign of water, as the Moon governs the tides that cause the waters to ebb and flow. The sign of Muluc also represents the power of purification that water can bring, from life-bringing rains to destructive floods. This is a reminder of the Mayan creation story that says the previous world was destroyed by water. Muluc teaches us not to resist when we need to go with the flow of life.

OC

Dreamspell **and English translation:** Dog
Keywords: Loves, Loyalty, Heart
Color: White
Direction: North

The unconditional love of the Dog is the key to this sign. The sensual nature of this sign requires mastery over animal instinct, but when the Dog is trained it becomes the most loyal and trustworthy companion. Oc represents the heart, but the Maya also associate the sign of Oc with misuse of sexuality and even promiscuousness. In Mayan mythology, it was the dog that brought humans the gift of fire.

CHUEN

Dreamspell **and English translation:** Monkey
Keywords: Plays, Magic, Illusion
Color: Blue/Black
Direction: West

Monkey is a mischievous and playful sign, the trickster of the bunch. Monkey days may not go as they are planned and can be full of unexpected surprises. The sign of Chuen loves to have fun at the apparent expense of others, but also teaches about the nature of illusion when we are taking ourselves too seriously. It is a lighthearted, humorous day sign.

EB

Dreamspell: Human; **English translation:** Road
Keywords: Influences, Free will, Wisdom
Color: Yellow
Direction: South

The power of free will, represented by the Human's ability to choose, is the key to this sign. Choosing wisely may serve the higher good. If not, the destructive foolishness of the selfish and self-centered man may prevail. Eb influences the unfolding drama of life through playing this key role. The sign of the Road suggests there are many turnings on the road of life to choose from.

BEN

Dreamspell: Sky Walker; **English translation:** Reed
Keywords: Explores, Space, Prophecy
Color: Red
Direction: East

The symbol of Sky Walker can be visualized as two pillars connecting heaven to earth. This is the role of this prophetic sign that is connected to Quetzalcoatl, the legendary One Reed and archetypal deity of this sign. Being the hierophant that connects these worlds allows the sign of Ben to be a conduit for prophecy and insight. The Maya associate the Reed with knowledge and authority.

IX

Dreamspell: Wizard; **English translation:** Jaguar
Keywords: Enchants, Timelessness, Receptivity
Color: White
Direction: North

This day sign, associated with the night sky, represents the power of timelessness. The ability to go into trance states to access knowledge from the spirit world is a key to this sign. The jaguar priests were the shamanic elite of the Maya, and Ix is their sign. Those born under this sign can step out of time to receive messages that can then be used for good.

MEN

Dreamspell **and English translation:** Eagle
Keywords: Creates, Vision, Mind
Color: Blue/Black
Direction: West

The Eagle's soaring perspective means it can see the bigger picture. Sharpness of vision and clear-sightedness are able to reveal the larger truth of a situation, transcending the more mundane, earth-bound outlook of everyday life. The Eagle's eye is a reminder of the planetary mind that modern society has lost touch with.

CIB

Dreamspell: Warrior; **English translation:** Vulture
Keywords: Questions, Intelligence, Fearlessness
Color: Yellow
Direction: South

The Warrior is forever questing in search of new understanding. This sign is determined and resolute, and proceeds where action is needed without fear. The result of this is that the sign of Cib may at times be perceived as being too serious. The Maya see Cib's associated animal of the Vulture as a largely positive bird, providing a necessary part of nature's recycling service.

CABAN

Dreamspell: Earth; **English translation:** Earthquake
Keywords: Evolves, Navigation, Synchronicity
Color: Red
Direction: East

Those born under this sign can use being connected to Earth to navigate through life. The Earth represents a matrix of synchronicity, which can be accessed by being attentive to the signs and willing to go with the flow without resistance. The sign of Caban is also strongly associated with the movement of Earth, especially earthquakes.

ETZNAB

Dreamspell: Mirror; **English translation:** Flint
Keywords: Reflects, Order, Endlessness
Color: White
Direction: North

The Mirror holds up a surface for us to see our own reflection. This image may be fragmented or appear broken, but in the Mirror's changing facets we have a tool that teaches us about our own endlessly changing natures. The sign of Etznab shows us there is not one true face but many, and that the order of things is constant change.

CAUAC

***Dreamspell* and English translation:**
Storm
Keywords: Catalyzes, Self-generation, Energy
Color: Blue/Black
Direction: West

The Storm is a completely self-contained ecosystem. It draws up the waters of the ocean and then transports them over the landmasses. It then drops the water, renewing the cycle. The sign of Cauac shows us how to conserve and nurture energy by becoming self-generating. Cauac is enormously powerful, catalyzing great change in its surroundings, but it renews itself from its own source.

AHAU

***Dreamspell* and English translations:**
Sun
Keywords: Enlightens, Fire, Life
Color: Yellow
Direction: South

Don't Ignore the Numbers

The number associated with the day sign is just as important as the sign itself. In fact, the qualities associated with these numbers are considered the more active or dynamic part of the equation, while the day signs are considered more passive or static. The traditional Maya believe some of the numbers are luckier than others. The higher numbers are especially auspicious when they appear during divination rituals. One is also a very important number, being associated with numerous mythological figures and gods including Quetzalcoatl, the hero twins, Hunab Ku, and the planet Venus.

The Sun is the final sign and is often considered the most auspicious and prophetic of them all: a day doubled. The Sun represents the culmination of the cycle of creation and its very source. The sign of Ahau, the symbol of the Mayan kings, enlightens us with its life-giving rays and governs the growth and fruition of all things. Ahau represents both the furnace of original creation and the brightness of our eventual enlightenment.

42. The Thirteen Numbers and Their Powers

The numbers begin with the Yucatec Mayan name for the number followed by the *Dreamspell* name of the tone and the corresponding keywords. The summary is intended to give an idea about how the numbers fit together in the *Dreamspell* system to create a progression that links the cycle of thirteen together.

Mayan Number: HUN (one)

Dreamspell Tone: Magnetic Tone
Keywords: Unify, Attract, Purpose
Description: The first tone represents unity and new beginnings. An intention set on this day will carry throughout the whole thirteen-day period that follows. Each day has its own energetic function that assists in the process of harmonious manifestation in tune with the fractal time wave of thirteen pulses.

Mayan Number: CA (two)

Dreamspell Tone: Lunar Tone of Challenge
Keywords: Polarize, Stabilize
Description: The function of the second day is to polarize the intention of the first day. This means looking at the challenges to the manifestation of the desired goal. Clearly identifying these obstacles before attempting to move forward is a key to using the wavespell process.

Mayan Number: OX (three)

Dreamspell Tone: Electric Tone of Service
Keywords: Activate, Bond
Description: After the original intention and its challenge have been assessed, the third stage is to activate and energize the process of achieving the goal. The question to ask at this stage is "What does the intention serve?" Answering this removes the personal ego from the equation by aligning with the greater good.

Mayan Number: CAN (four)

Dreamspell Tone: Self-Existing Tone of Form
Keywords: Define, Measure
Description: At the fourth tone, the four corners of a square can be represented. This is the first point that the intention can begin to be defined in the third dimension. A detailed plan of action is required so that form can be given. Write down exactly what you plan to do.

Mayan Number: HO (five)

Dreamspell Tone: Overtone Tone of Radiance
Keywords: Empower, Command
Description: After the action plan is drawn up, the completed blueprint is now empowered by drawing in the necessary resources to accomplish it. The fifth tone represents the point to focus on gathering what is required. Being in command means taking control of the process by drawing in whatever is necessary.

Mayan Number: UAC (six)

Dreamspell Tone: Rhythmic Tone of Equality
Keywords: Organize, Balance
Description: Having gathered resources, the sixth part of the thirteen stages is to effectively organize all aspects of the project. Balance is achieved by placing all the elements into the appropriate place relative to each other. Find the order in apparent chaos.

Mayan Number: UC (seven)

Dreamspell Tone: Resonant Tone of Attunement
Keywords: Channel, Inspire
Description: The seventh step of this process is the midpoint. This is a point at which the space can be created for divine inspiration to be accessed. At this pause, opening up to a higher source may yield insight. Inspiration is often unexpected, but by being open to it we encourage the best possibility of recognizing it when it does happen.

Mayan Number: VAXAC (eight)

Dreamspell Tone: Galactic Tone of Integrity
Keywords: Harmonize, Model
Description: The eighth stage is about testing the integrity of what has been brought together so far. This is an opportunity to try out various solutions in a dry run to explore different possible results. This experimentation and modeling ensures that when the final action of manifestation is taken, the project has the best possible chance of perfect success.

Mayan Number: BOLON (nine)

Dreamspell Tone: Solar Tone of Intention
Keywords: Pulse, Realize
Description: Upon reaching nine, it is now time to fully engage the pulse of intention to realize the goal that has been set. This takes what has been assembled so far and expresses it with full and complete commitment. Acting without hesitation or second-guessing sends a clear message to the receptive universe. This allows the law of attraction to be accessed in an optimal way for the best possibility of success.

Mayan Number: LAHUN (ten)

Dreamspell Tone: Planetary Tone of Manifestation
Keywords: Perfect, Produce
Description: The tenth tone is the point at which manifestation occurs. By having undertaken the careful preparation in the previous nine steps, all that remains is to focus on perfection. The next three tones will concentrate on recirculating the energy that has been created in this process so that on the first day of the next cycle, everything is ready to begin again.

Mayan Number: HUN LAHUN (eleven)

Dreamspell Tone: Spectral Tone of Liberation
Keywords: Dissolve, Release
Description: The spectral tone is about completely releasing what has been created. This is a necessary part of creating in tune with nature, which is always recycling what it has created. This step allows the dissolution of attachment, avoiding the trap of being drawn out of the present moment by lingering on the completed project.

Mayan Number: CA LAHUN (twelve)

Dreamspell Tone: Crystal Tone of Cooperation
Keywords: Dedicate, Universalize
Description: The twelfth stage is for sharing what we have experienced with others. In the thirteen-day cycle of the wavespell, this is marked by a crystal day gathering, where those who are working with the calendar in this way can meet and share what they have learned. This sharing is an important part of the process and prepares the way for beginning again.

Mayan Number: OX LAHUN (thirteen)

Dreamspell Tone: Cosmic Tone of Presence
Keywords: Endure, Transcend
Description: The last of the thirteen is a day of meditative awareness, where being in the present moment is the sole focus. These last three stages are what most significantly distinguish the wavespell from other forms of working with intention. Manifestation is only part of the holistic process anchored in the natural time of the here and now. The intended product of following these cycles is personal development and increased awareness.

PART THREE

Changes on Earth and What They Could Mean for 2012

Phrases like "timewave zero" and "galactic superwave" may sound like they come straight from the scripts of sci-fi movies. It's true that many of the concepts associated with 2012 are difficult to grasp, and that leads laypeople to criticize them as far-fetched.

In this section, you'll read about everything from cosmic rays to pole shifts; some phenomena have been proven true while others are just hypotheses. Ultimately, you must decide what you believe based on the information available.

43. What Does December 21, 2012 Have to Do with the Winter Solstice?

The researcher and author John Major Jenkins made one of the most interesting modern discoveries about the Mayan calendar's end date of December 21, 2012. His book *Maya Cosmogenesis 2012* puts forward the theory that the calendar end date also points to what will be a very significant astronomical event: the rising of the winter solstice sun in conjunction with the center of the galaxy. This is an event that only happens once every 25,771 years.

The alignment of the winter solstice sunrise with the center of the galaxy occurs because of something called the precession of the equinoxes. Precession refers to a change in the direction of the axis of a rotating object. From Earth's point of view, the sunrise on any fixed day will rise against a background of stars that changes over time.

Precession is a slow process in which the stellar background against which the sun rises moves approxi-

Equinox

An equinox occurs when the sun is located vertically over the equator. This happens twice a year. The first one, on March 20 or 21, is the spring or vernal equinox for the northern hemisphere and the autumn equinox for the southern hemisphere. The second equinox takes place on September 22 or 23.

Precession of the equinoxes

A gradual shift in the orientation of Earth's axis of rotation that traces out a conical shape in a cycle of approximately 25,771 years.

mately 1° every seventy-two years. This also creates what is known as the changing ages of the zodiac in astrology. The stellar background for sunrises and sunsets is always found in this 14° wide belt called the ecliptic.

The ecliptic is the path the sun follows as it moves through the year. Each day, the location of the rising sun appears to move against the background of the stars as it gradually moves its way around the circle of the ecliptic. In a year, the location of the sunrise will have returned to almost exactly the same position in the sky. The difference will be around 0.0138°. Measuring this difference is how we detect precession. Over the whole 25,771-year cycle, these accumulated differences will mean

Ecliptic

An imaginary circle projected into the sky from the plane of the solar system.

that the rising point of the sun, measured on the spring equinox, will have completed a 360° circle and will have returned once more to exactly the same place in the sky.

The ecliptic is then divided into twelve constellations, which form the signs of the zodiac. Together, the twelve signs create a wheel of the year. The cycle of precession, where the rising and setting points of the sun gradually move against the solar year, means that the zodiac can also be used to mark a greater wheel of the ages. Each of these zodiac ages lasts 2,160 years, during which the rising sun at the vernal equinox will stay within one of the signs. Twelve zodiacal ages make one great year of 25,920 years, a close approximation of one precessional cycle.

44. The Winter Solstice Galactic Alignment Actually Lasts 38 Years

The conjunction of the winter solstice with the galactic equator is not just a one-time event that will happen on December 21, 2012. The disc of the sun is one half a degree wide, so it will take about thirty-six years to fully cross the galactic mid-plane. This means that the conjunction is happening now and will be happening for some years after 2012. In fact, the period of conjunction began around 1980 and will continue until approximately 2018.

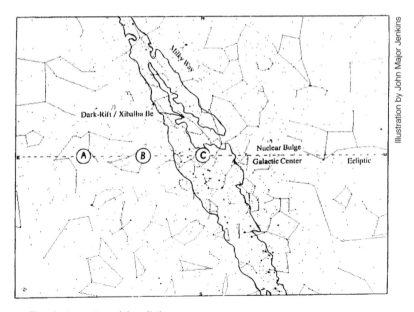

Illustration by John Major Jenkins

▲ The galactic equator and the ecliptic

The alignment is formed by the location of the winter solstice sunrise crossing the galactic equator. The galactic equator can be visualized as an imaginary circle projected onto the sky in a similar way as the ecliptic but in the plane of the galaxy. The circles of the galactic equator and the ecliptic intersect at an angle of 60° to each other. This creates the effect of a cross in the sky.

The classic Maya called this cosmic cross the Sacred Tree, and it was extremely important to them. Jenkins believes that this is the real origin of the cosmic Mayan world tree depicted on the sarcophagus lid of Lord Pacal. It also corresponds to the cahib xalcat be or four junction roads, the crossroads of Mayan mythology recorded in the Quiché book of the Popol Vuh. Finally, this is the site of the entrance to the underworld, Xibalba.

Representations of the four roads also appear in the layout of a typical Mayan village. Jenkins theorizes that these—as well as the ceremonial centers

An Optical Alignment

The galactic alignment of the 2012 era is an optical one, not a physical one. We are not actually crossing the center of the galaxy. It is just that a neat conjunction makes it look like we are, a bit like the hands on a giant clock aligning to cosmic midnight. It is this that Jenkins is bringing attention to. His research suggests that a pre-Mayan culture found at the site of Izapa knew about this conjunction and actually went as far as to align the building of their ceremonial sites to mark the event. This was an event that, for them, was still nearly 2,000 years in the future.

built by the Itzá and the Maya—are laid out to reflect this cosmological order. The center axis of the city would be a copy of the celestial order above. The celestial crossroads or cosmic navel represent the place of creation in the Mayan cosmology. It is from this location that all creation emanates; it is the absolute center of the Mayan universe.

45. Other Galactic Cycles to Know

Some writers have confused the idea of galactic alignment with another galactic cycle. This is the much, much longer journey our solar system takes as it moves above and below the central plane of the galaxy. This up-and-down motion happens at the same time as our very slow orbit around the center of the galaxy. Our star, the Sun, is located in one of the arms (sometimes called the Orion arm) that radiate out in a spiral from the center of our galaxy. Our Sun travels through and between these spiral arms as it revolves slowly around the galactic center in an orbit that takes about 226 million years.

This means that in the entire lifetime of our planet, we have only orbited the galactic center about twenty times. At the same time, our Sun also moves up and down in an oscillation toward and away from the galactic equator. This is sometimes confused with the winter solstice/galactic equator alignment of 2012, but this cycle takes much longer, passing through the galactic mid-plane about every 35 million years. In this long cycle, the sun is actually moving away from the galactic equator at this point in time. It will not cross it again for approximately 30 million years.

At the moment, our solar system is moving toward the more densely populated center of our local galactic arm. This means we are entering an area that has more red giant stars. These can be dangerous because they can explode into what are known as type II supernovae. The solar system is currently inside a sixty-light-year-wide interstellar bubble that a previous type II supernova explosion has largely cleared of interstellar dust particles. This means our night skies are clearer than they otherwise would be.

46. The Black Road to the Underworld

There is a dark patch in the Milky Way close to the galactic center often called the dark rift. This is caused by an interstellar dust cloud obscuring the light coming from the stars at the core of the galaxy. The Maya believe that this feature marks the black road to the underworld. They imagine it as a celestial mouth, from which deities and kings are born. It is from this dark rift that the winter solstice sun will appear to emerge when it rises on December 21, 2012. The appearance of the sun from the dark rift would be a natural point in the Mayan cosmology from which to measure a beginning or ending point.

The rising of the winter solstice sun was a significant event to the Maya, and it is recorded as the most important of the four quarters of the year. The winter solstice is still the focus of a

Jenkins's Interpretation of a Mayan Myth

Jenkins takes a myth found in the Quiché Mayan book *Popol Vuh* and applies its symbolism to the Mayan calendar. In this story, the First Father, One Hunahpu, represents the birth of a world age. During his journey to the underworld he is killed by the lords of Xibalba. His skull comes to rest in a crevice in a tree, which Jenkins equates to the cosmic world tree of the galaxy. The crevice represents the dark rift or the cosmic birth canal of the galactic cosmic mother. The skull spits into the hand of Blood Moon, who bears two sons, the hero twins. They defeat the lords of Xibalba, allowing One Hunahpu to be reborn. During December 21, 2012, the winter solstice sun is "birthed" from the dark rift. The name One Hunahpu is also equivalent to the calendar date One Ahau. This could also be taken to mean First Sun, and One Hunahpu represents the winter solstice sun.

midnight ceremony held by the indigenous Maya to mark this event. This belief is also reflected in other ancient cultures. For example, in the Neolithic temple of Stonehenge, the trilithon of stones aligned to the winter solstice is the tallest. There is also a myth in the Celtic traditions of the king—a common symbol for the sun—dying the day before the winter solstice, known as the kingless day, to be reborn anew the day after.

47. How Izapa Ball Courts Relate to 2012

The site of Izapa is located in southern Mexico, near the Pacific coast and to the west of the Guatemalan border. This is a likely birthplace of the Long Count calendar and the site of some very interesting astronomical alignments, one of which seems to point to 2012. Izapa was populated between 400 B.C. and A.D. 100 and was the dominant site in its region. Some of the very earliest Long Count inscriptions, dating back to around 40

B.C., have been found in locations close to Izapa, and would almost certainly have been influenced by it. Unfortunately, no Long Count inscriptions have yet been found at Izapa itself.

In *Maya Cosmogenesis 2012*, Jenkins describes many alignments from this site that relate to different astronomical cycles. The most important of these for its relationship to 2012 is the alignment of the ceremonial ball court. Ball courts are very common features in Mayan ceremonial sites, and ball games, played with a small ball made of rubber, are common throughout Meso-American indigenous culture. Like most other aspects of Mayan life, the symbology of the ball court and the ball game has cosmological roots.

The game is played with two teams competing to put a small rubber ball through a hoop mounted high at the end of the court, using only their knees and elbows. The rest of the court consists of two parallel banks of terraces used for spectator seating.

Murals found on the great ball court of Chichén Itzá show the primary ball player being decapitated. His head then becomes the game ball. This has given rise to the idea that the captain of one side was sacrificed after a ball game. Jenkins believes this mural actually refers to a high deity and is metaphorical.

Jenkins's theory is that the game ball represents the winter solstice sun and the goal ring into which it must be placed represents the dark rift of the Milky Way, which marks the entrance to the underworld, Xibalba. This suggests that the ball game represents knowledge of the cycle of precession, and also points toward the calendar end date of December 21, 2012. The ball court at Izapa

Thinking Ahead

At the time the ball court was built, the rising point of the sun was still approximately 30° below the galactic equator. If the symbolism of the ball game and the galactic center is correct, the ball court could be seen as a sort of celestial alarm clock pointing at the conjunction happening in our era. Being able to mark an astronomical event like this, 2,000 years before it happens, is remarkable and shows how important the conjunction was to the Izapans. In effect, they encoded their knowledge of this galactic cosmology into this monument. Their vision of this event was already complete before the rise of the classic Maya some hundreds of years later.

is aligned east-west with the goal ring in the east. Looking down the court toward the goal, the winter solstice sun could be seen to rise in the background.

48. Mayan Creation Focuses on the *End* of Certain Time Periods

A conjunction of a solstice or an equinox sunrise and the galactic equator will happen once every 6,442 years. The winter solstice conjunction happens only once every 25,771 years. What makes this even more significant is that the winter solstice conjunction is also the end of a 5,125-year cycle of the Long Count. If you multiply the length of the Long Count by five, you get a total of 25,625 years, very close indeed to one cycle of precession.

Jenkins suggests that each of these five multipliers could correspond to one of the Mayan five worlds of creation. The Long Count would then be just one-fifth of an even bigger precessional calendar. This neatly ties up the five worlds of creation found in the *Popol Vuh*, with the conjunction of the place of creation and the sun on December 21, 2012. These events happening together, he suggests, perfectly mark the creation event at the ending of a whole precessional cycle.

The Maya attribute creation and birth to the *end* of a cycle, rather than the beginning. For example, the birth of a child follows its gestation period of approximately 260 days, or one Tzolkin. Each of the Mayan time cycles was named after its last day, not its first. In Mayan prophecy, it was the last day that determined the oracle for the time period.

Cycles that finish on an Ahau day are of particular significance. This happens in the 260-day Tzolkin, at the end of every twenty-year katun, and the end of the thirteen baktuns on December 21, 2012, which is the day 4 Ahau. According to Jenkins, this is the day when the New Sun emerges from the cave of creation. This is a date that marks the shifting of the world ages and a cosmic re-creation. From this point of view, the thirteen baktuns of the Long Count could be seen as a planetary pregnancy leading to the birth of something entirely new.

By most estimates, precession would take about 200 years of direct observation to calculate, so the Maya may have discovered it as early as the time of the first Long Count inscriptions. Why they did not add their knowledge of the cycle to their calendar may be explained by the fact that they encoded it into the doctrine of the five worlds of creation, although this cannot be proven.

49. What Psychedelics Predict for 2012

Another theory about 2012 also focuses on the end date of the Mayan calendar, but the inspiration for this hypothesis comes from a very different source from those we've looked at so far. The brothers Terence and Dennis McKenna were on a journey to the Amazon to research indigenous psychoactive plants, when an extended psychedelic experience led them to a new interpretation of the ancient Chinese oracle of the *I Ching*. This led them to predict a remarkable event for December 2012.

Cultures in central and southern America have a long history of the use of psychoactive and psychedelic plants for the purposes of divination and healing. The Amazonian drink ayahuasca, for example, goes back at least 2,000 years. Made from a combination of two or more plants that are only moderately psychoactive when taken alone, the ayahuasca brew contains a brain chemical called dimeythltryptamine, or DMT. DMT in this form is a powerful psychedelic. Users report visions that often feature animals representing spirits in experiences that can last between four and eight hours. Ayahuasca is widely used by a number of indigenous tribes throughout the Amazon as a sacred plant medicine. It is respected for its potency and is used to heal and also to relate to the spirits of the rainforest. It has also been used for sorcery.

It was ayahuasca that the McKenna brothers, who are both ethnobotanists, were searching for on their Amazonian expedition. They were particularly interested in studying a form of auditory hallucination that had been reported by users of the brew. This led them to create an experiment in which they took ayahuasca in combination with the psychoactive mushroom *Stropharia cubensis*. The result was a long and potent visionary experience that is detailed in the book *True Hallucinations*.

This experience inspired Terence McKenna to develop his timewave zero hypothesis. Like Calleman's pyramid model of the calendar, timewave zero also predicts an ever-greater acceleration of time toward the Mayan calendar end date of December 21, 2012. In this case, the level of acceleration actually becomes exponential. If the timewave hypothesis is correct, we are heading toward an event in 2012 that will result in the creation of a new sense of time, or an end to time as we know it.

50. McKenna Theorizes that the *I Ching* Is a Lunar Calendar

Terence McKenna's timewave zero hypothesis is constructed around one single fact about the *I Ching*: its numerical

Jenkins's Thoughts on Psychedelic Plants

Jenkins speculates that psychedelic plants could have led Mayan kings and priests to the discovery of the galactic alignment. Many Mesoamerican tribes, the classic Maya among them, have a history of using psychoactive plants as part of their culture. These plants include datura, morning glory, tobacco, and psychedelic mushrooms. Jenkins points to the availability of many plants that contain DMT in the Maya's bioregion. Jenkins thinks the Maya gained their knowledge of galactic cosmology by taking DMT and going on visionary shamanic journeys to the galactic center. The Mayan understanding of the galaxy was not just a product of reasoning or observation, but direct knowledge gained through shamanism. This would certainly help explain the development of the Maya's stunning calendar and astronomical sciences so early in their history at Izapa.

foundation. The basic unit of the *I Ching* is the line, or yao. Six of these make a kua or hexagram, and there are sixty-four different permutations of these kua that make up the complete oracle. This means the whole system is based upon six multiplied by sixty-four, making a total of 384 lines.

The timewave theory came about when McKenna linked this number to an important lunar cycle. Each lunation of Earth's moon is approximately 29.53 days, and a lunar year of thirteen lunations totals approximately 383.89 days, a figure very close to 384 days. This, McKenna suggests, means the *I Ching* might, in fact, be an ancient lunar calendar.

It is widely accepted that the oldest calendars kept by the Chinese, going back to Neolithic times, were lunar. A calendar based on 384 days or thirteen lunations is a good candidate for basing a lunar year upon because it is just 2.4 hours short of a complete day. By adding one day every ten lunar years, this calendar will only lose one day every 454.5 years, making it very accurate.

The sixty-four hexagrams of the *I Ching* can be ordered in different ways, but one of the most ancient orders is called the King Wen sequence (named after an ancient Chinese emperor). McKenna describes this as "the oldest preserved human abstract sequence" because its origins date to the beginning of known history. The hexagrams in the King Wen sequence are placed into pairs, where the second hexagram of a pair is obtained by inverting the first. This series of thirty-two pairs exhibits some very unusual numerical properties that McKenna used to form the basis of timewave zero's mathematics.

McKenna was eager to find the intellectual principle that lies behind the ordering of the thirty-two pairs of *I Ching* hexagrams, which was unknown. He theorized that the structure could be expressed as a first order difference between the hexagrams. This is the number of lines that change between one hexagram and the next. It will always be an integer between one and six. What he found was that an exact 3:1 ratio of odd to even transitions was maintained in the King Wen, and that the number five was excluded completely. This seemed to be a profoundly unusual structure.

51. McKenna Links an *I Ching* Waveform to the Space-Time Continuum

McKenna was also struck by the fact that the King Wen sequence contained a unique aspect: The first and last three positions were mirror images of each other. This means that if you take the whole sequence, you can reverse it and place it over the original and each hexagram will be paired with its opposite. This backward and forward combination of the first order differences creates closure at the beginning and end of the sequence when it is mapped this way. This creates what can be seen as a single waveform.

Waveform

The shape and form of a graph plotting certain points.

McKenna concluded that the waveform made from the whole sequence of the *I Ching* could be used as the basic unit to form the hierarchy of a multilevel calendar. Treating the entire sequence as one unit, or yao, he further multiplied the wave by six and sixty-four. This creates different layers, still based on the mathematics of the *I Ching*, which can be used to map every possible level of time from the very large to the very small. A lunar year of 384 days multiplied by sixty-four gives a period of around 67.3 years. This can then be multiplied by sixty-four again, giving a period of 4,306 years. This period is very close to the length of two zodiacal ages—4,320 years. Six of these periods actually give a closer approximation to one cycle of precession than the great year of astrology. This led McKenna to believe that the relationship between the *I Ching* and the lunar calendar was fundamentally harmonic and could be used as a key to understand the whole unfolding process of history.

By such expansion through multiplication, McKenna was able to apply this highly complex waveform to the whole of the space-time continuum. To achieve this requires twenty-six different levels of the original *I Ching* wave, each of which is a factor of sixty-four larger than the preceding level. The hierarchy of timewave zero extends in this way to cover a period of 72.25 billion years at the large end, all the way down to the subatomic range of Planck's constant on the microscale. At this point, further subdivisions on the quantum level are no longer meaningful.

52. Novelty Theory

McKenna's timewave can be read as an index of how much of what is occurring at any given time is novel or new. The more unexpected or original, the more novelty the event has. A very novel event could be the invention of the telephone or the Asian tsunami of 2004. The timewave doesn't tell you specifically what will happen or where, but it does describe how novel it is. By looking at the repeating patterns and their corresponding novelty, the idea is that we are able to predict *when* interesting events will occur but not necessarily *what* will be happening.

The accelerating changes in our world over the last few decades contain many good examples of novelty. From the invention of the automobile to the popularity of Facebook, new social and technological forms are appearing more and more rapidly. Change is increasing, and this increase is predicted by the timewave. As a model of time, it embraces the spirit of the modern world far better than the arrow of linear time, which has its origins in the medieval world of Christian theology.

The timewave suggests that different periods of history have resonance

Merging the Timewave with History

McKenna uses the bombing of Hiroshima as the historical anchor point for the timewave. From this single point, the timewave is projected as an overlay onto history. After this key point, it takes the timewave exactly one period of 67.3 years to reach closure. This made the original end date of the timewave November 17, 2012. When McKenna discovered that the end date of the Mayan calendar was just a few weeks later, he simply changed the end point to fit this new insight. This means the original anchor point is somewhat lost, as the whole wave moves forward by twenty-eight days. There is no definitive or empirical reason why the timewave should finish on December 21, 2012. For McKenna, the end of the Mayan calendar just seems to fit in well with the sort of unprecedented event the timewave is predicting.

with each other. For example, some commentators suggest that the timewave for the period of World War II, 1939–1945, closely resembles that for the period 2007–2012. Both are characterized by a major spike toward the end of the period. In the case of World War II, McKenna equates this with the detonation of the first atomic bomb at Hiroshima. In the case of 2012, events could be even more dramatic.

53. The Timewave Predictions for 2012

What the timewave predicts for its end point at 6:00 A.M. on December 21, 2012 is quite extraordinary. Whereas Calleman and Argüellés are predicting major shifts in consciousness, McKenna's theory predicts events so strange they are pretty much unimaginable.

The timewave predicts that changes that have as great an impact as the invention of agriculture or the Industrial Revolution will happen no fewer than eighteen times in 2012—and that's only on the last day. In fact, there will be as much change in the last part of that year as was contained in half the previous life span of the universe, from the big bang to the birth of the sun. This mind-boggling acceleration will increase exponentially. Five of these changes will occur during the year, eighteen in the last day, and thirteen in the last second of the timewave.

This incredible set of circumstances represents a drawing together of multiple streams of being merged into one unified thing. This is the singularity that waits for us at the end of time.

McKenna calls this the eschaton, a sort of strange attractor that is drawing the evolution of the universe toward itself. His ideas represent a totally different form of time that has no past or future; it is comparable only to a state of revelation. This is the fractal at the end of time—a vortex of change that will completely transform the human experience.

In his rationale for the extraordinary compression sequence that ends the timewave, McKenna observes that the laws of physics themselves seem to vary by their scale in space. Modern science is still working to reconcile the very different laws that seem to operate on the quantum scale and at the cosmological ones. He suggests that these laws not only vary by the scale of space but also by the scale of time.

As we approach 2012, the fractal scale of the timewave diminishes the closer we get to the end point. As the scale diminishes, McKenna predicts that the physical laws that accompany them will change. Hence, the possibility of huge amounts of innovation and novelty as we get closer to this event, much like a big bang in reverse.

Eschaton

The end of everything, especially time; the final destiny of the world.

What McKenna has done with the timewave is paint a picture of the possible, rather than the probable. First published in 1975, the timewave is actually one of the earliest attempts at explaining what might happen in 2012, and it is still one of the most original. By McKenna's calculations, we are about to be sucked into a whirlpool of novelty from which there will be no return. This event may be caused by the collision of two universes, causing all matter and antimatter to be totally destroyed. This could represent the liberation of all beings from the burden of time, as we would emerge from the 2012 event into a new universe composed entirely of light. What this experience would actually be like is beyond even McKenna's vivid imaginings.

54. Critiques of the Timewave

Critics of McKenna's theory point out that his anchoring points into history are arbitrary and subject to debate. If you change the anchor point, you change the end date. Nonetheless, since the time-wave reaches closure very quickly after the anchor point, it would have to be sometime in our modern era.

In addition, skeptics say the nature of novelty is subjective rather than objective. What might seem to be particularly novel for one person may seem less so for another, making the timewave unscientific. In the case of recent events that are generally agreed to be novel and to have wide significance, the timewave hypothesis has not been particularly successful in making clear predictions. For example, the events of 9/11 barely scored more than a minor deviation from most of the projections of the novelty curve.

The timewave, for all of its cosmological predictions, actually represents a view of time that is centered on humans. McKenna made no apology for this. He believed that in the hierarchy of time, human perception was exactly in the middle order. In other words, we are not cosmically insignificant, as previously assumed by most science, and our ability to consciously perceive the universe is an important part of cosmic order. The worldview of the timewave is consciousness-centered and the existence of an objective physical universe, apart from its perceiver, doesn't figure. McKenna speculates that the world itself may come to be seen as a "fluctuation of a vacuum domain, albeit a long-lasting one," a spontaneously arising artifact of consciousness itself.

As a hypothesis, timewave zero has some very compelling qualities. It challenges the idea of linear time in a more radical way than has ever been done before. It also provides another useful model that points to a possible cause for the feeling of acceleration of time that many people are experiencing. It does these things with many brilliant insights about the relationship between the *I Ching*, DNA, and the nature of time itself. In many ways, the timewave is the quintessential 2012 theory, containing dramatic consciousness-expanding revelations with a mind-blowing psychedelic apocalypse at the end. It is a big idea. The timewave doesn't just predict the end of history; it predicts the end of time.

Where it appears more flawed and limited is when it is taken literally as a predictive model. Novelty is open to interpretation. Different anchor points create different timewaves. Nor is there any way to subject the hypothesis to scientific proof. If the claims contained in the timewave theory had only attempted to explain the span of *human* history, perhaps it would be more manageable and easier to accept.

55. The Possibility of a Galactic Superwave

One of the more devastating possibilities regarding 2012 comes from a theory that the galactic alignment the Maya may have been targeting could also be the due date for a major cosmic inundation. Astronomers have noted that between one in five and one in seven spiral galaxies appear to be in the process of exploding. The core nuclei of these galaxies can often be seen to glow as brightly as the whole galaxy itself. This can completely mask the features of the spiral arms in telescope images. These violent eruptions are huge in scale and can become more than 100,000 times brighter than our own galactic core. These actively exploding spiral galaxies have been called Seyfert galaxies, after their discoverer, the astronomer Carl Seyfert.

According to astrophysicist Paul LaViolette in his groundbreaking book *Earth Under Fire*, astronomers are now realizing that these explosions are not confined to a particular type of galaxy, but occur in all spiral galaxies on a periodic basis. About 80 to 85 percent of the time a spiral galaxy will appear normal, but in the remaining 10 to 15 percent they are actively erupting. These explosions radiate out from their galactic cores and, in some cases, can engulf the entire galaxy. The length of the explosions can last from hundreds to several thousands of years.

If one of these galactic superwaves was heading in our direction, we wouldn't even be able to see it. Light takes thousands of years to reach us from the center of the galaxy, so we probably wouldn't know about the superwave until it arrived. The consequences for our solar system could be far reaching, and for life on Earth, possibly severe. The superwave would bring with it a huge amount of interstellar matter. This would then be attracted into the gravitational field of the sun and the dust would form a shroud around our star.

To understand how this might affect our star, we can look at the characteristics of the T Tauri class of star. These have substantial dust clouds around their equatorial planes. The dust cloud causes the T Tauri stars to behave quite differently than our sun. These stars flare intensely, anywhere from between 100 to 1,000 times the amount our sun does. They also put out a much greater solar wind, light, and ultraviolet radiation than the sun does, although if the dust cloud were absent, the star itself would be of a similar size.

If our sun started to behave like a T Tauri star, the increase in solar activity would be dramatic. Average temperatures would increase substantially, probably causing massive flooding from rapidly melting ice and unbearable temperatures around the equator. In addition, the direct effects of the flares themselves could wipe out global communications systems or create effects similar to a massive airborne nuclear explosion. A reasonably sized galactic superwave event could be responsible for triggering an extinction cycle.

56. Predicting a Superwave by Analyzing Supernovae

If we can't see the superwave coming, how can we know if one is likely to arrive? LaViolette suggests that we may be able to detect the motion of a galactic superwave by looking for a pattern in supernova explosions. He theorizes that as the superwave passes, explosions of the unstable blue supergiant stars that are prone to this kind of explosion might be triggered by the gravitational tide of the wave. By looking for a pattern in supernova explosions, we might be able to detect the aftereffects of a passing galactic superwave as it moves through the galactic arm.

In analyzing the dates when recent supernovae have been observed, LaViolette shows that it is possible that all four supernovae recorded within historical times may have been triggered by the same galactic superwave. Each of these supernovae occurred at a different distance to our planet and was observed on a different date, but by analyzing and comparing these times and distances it is possible to tell if it was likely that they may have been triggered by the same superwave event horizon. For example, the Crab Nebula supernova explosion was recorded by the Chinese in A.D. 1054 and was clearly visible to the naked eye at that time. The remnant of this event is the famous Nebula and is approximately 6,585 light years away. LaViolette calculates that if a superwave triggered this supernova, it would have passed our solar system about 14,000 years ago. The three other significant supernovae of recent times—Cassiopeia A, Tycho, and Vela XYZ—also fit closely into this model. All four correspond to a possible galactic superwave trigger that would have passed our solar system about 13,400 years ago.

This period of 13,400 years closely corresponds with the last major mammalian extinction on Earth. It is also very close to one-half of a precessional cycle ago. LaViolette's theory is that galactic superwaves pass us once every precessional cycle. There is also the possibility, he believes, of a 13,000-year recurrence interval. If that proves to be true, it means that a significant superwave event could be imminent. Conventional scientific opinion suggests that superwave events are much less frequent than this, but LaViolette presents some interesting evidence to back up his claim.

Scientists have developed a technique for estimating the amount of cosmic-ray particle radiation that the planet is exposed to. When cosmic radiation hits Earth's atmosphere, the rays react to create tiny quantities of a rare radioactive isotope called beryllium-10, which then falls to Earth. By taking samples of ice cores at the polar ice caps, scientists are able to see how much beryllium-10 is concentrated in the layers that are formed each year. This can be used to interpret how much exposure there has been to cosmic radiation. There are cosmic-ray peaks that appear around 14,150 years ago; 36,800 years ago; 60,500 years ago; 89,500 years ago; and 103,500 years ago. This pattern corresponds quite closely to the length of the precession of the equinoxes. LaViolette believes that passing galactic superwaves could have created these cosmic ray influxes and suggests that there is a definite and predictable relationship between the occurrence of superwaves and Earth's precession.

57. The Great Flood and Ice Melt

Flood stories appear in the mythologies of more than ninety different cultures on all of the populated continents of the planet. One flood appears in the Bible as the story of Noah. The Maya believe that the last world was destroyed by a great flood. They called the flood Hun Yecil, meaning the "inundation of the trees," because the deluge happened so quickly that all the trees were swept away with it.

One of the most famous flood stories is that of Atlantis. The original story of Atlantis and the flood was told in Plato's *Timaeus* dialogue. LaViolette believes the story is actually an allegory and wasn't intended to be read literally. The clue for this is that Poseidon (Neptune) was given all the seas as his dominion as well as Atlantis, which was his only land territory. LaViolette's theory is that Atlantis wasn't a continent at all but the vast ice sheet that once covered North America. Hence, being water, it was really still part of Poseidon's domain. The myth of Atlantis wasn't intended to be read as the sinking of a continent but as the melting of the ice sheet, which was responsible for the great flood. The timing of the sinking of Atlantis is around 9600 B.C., which coincides with the climatic warming that occurred at the end of the last Ice Age, known as the Younger Dryas. According to LaViolette, the Atlantis flood was triggered by the large amount of cosmic dust injected into the solar system by the superwave event that happened 2,000 years previously. This dust altered the sun's radiance, resulting in a substantial change of climate on Earth.

The dialogues also refer to two previous "attacks" of Atlantis upon Greece, which in LaViolette's interpretation would be two previous floods. These correspond to the glacier wave flooding associated with the superwave that happened 14,000 years ago and the species extinction peak of 12,700 years ago.

LaViolette's view is that the ancient Greeks had encoded a sophisticated astronomical knowledge in these myths. For example, it appears that the cycle of precession is encoded in the Great Year of the Greeks that lasted 26,000 years. It consisted of two epochs, both of which ended in widespread disaster:

- The great summer or ekpyrauses lasted for half the cycle and ended in combustion.
- The great winter or kataclysmos lasted the other half and ended in deluge.

LaViolette estimates this deluge phase corresponds to the point at which precession will have brought the galactic center to its southernmost point. This will occur approximately 270 years from now, so very close to our epoch. The orientation of our planet to the galactic center also means that any galactic superwave that does arrive is likely to cause maximum damage.

58. Superluminal Gravitational Waves and the Asian Tsunami

In his 1983 doctoral dissertation, LaViolette suggests that the arrival of a galactic superwave event may be preceded by a gravitational wave. Gravitational waves are emitted by neutron stars, black holes, and probably the galactic center. They have two unique properties:

1. They don't need any type of matter to be propagated.
2. They can pass through any intervening matter without being dissipated. This makes them different from electromagnetic radiation, which may be blocked out by interstellar dust.

The initial intensity of a galactic nuclei explosion may cause the gravitational wave to actually reach superluminal speeds. This could be as much as sixty-four times the speed of light, giving the wave a head start on the electromagnetic component of the blast. LaViolette has predicted that a superwave impact may be experienced as a gravitational wave impact first, followed by a major gamma-ray burst.

The massive earthquake that caused the Asian tsunami of 2004 was followed just forty-four hours later by an enormous gamma-ray burst. The magnitude of both of these events was extremely unusual. The earthquake on December 26 measured 9.3 on the Richter scale. That's ten times bigger than any other earthquake in the last twenty-five years. The gamma-ray burst that followed on December 27 was the largest ever recorded by a factor of 100. The outburst was so powerful it released more energy in a tenth of a second than the sun emits in 100,000 years. The gamma-ray outburst came from a star called SGR 1806-20, located about 10° northeast of the galactic center at about the same distance to the center of the galaxy as our Sun, approximately 26,000 light years. The star is just twenty miles wide, but its weight is 150 times the mass of the sun. During its active phase, it was 40 million times brighter than our star.

Both the earthquake and the gamma-ray burst are categorized as Class 1 events. The probability of two such events falling so closely together is in the order of 5,000 to one against. Given this, LaViolette suggests that the Asian tsunami may have been caused by a gravitational wave impact. It is unfortunate that not one of the three major gravitational wave-detecting telescopes was online at the time of the tsunami, which means that the gravitational wave hypothesis can be neither confirmed nor disproved for this event.

These two very unusual events happening in close proximity almost exactly fit LaViolette's description of a superwave impact. The gamma-ray burst from SGR 1806-20 was by far the largest ever recorded, but a galactic nucleus

The Possibility of a Pole Shift

The idea that a pole shift will happen in 2012 is a popular one. LaViolette's research does point to a link between geocosmic cycles and magnetic field changes. The last really major disruption to Earth's magnetic poles happened around 12,700 years ago, about the same time as the proposed superwave impact. As a result, Earth's north magnetic pole moved to a location in the equatorial mid-Pacific for between ten and fifty years. This event was called the Gothenburg magnetic flip after the city it was discovered in. During this period, Earth's magnetic field fluctuated massively in step with the eleven-year sunspot cycle. This happens to a much smaller extent in a typical sunspot cycle, but during this period the peaks were hundreds of times more intense, approaching the levels found in T Tauri stars.

explosion could possibly contain more than 100,000 times even this vast amount of energy.

59. The Correlation Between Precession and Superwaves and 2012

It seems remarkable that both Jenkins's galactic alignment theory and LaViolette's galactic superwave theory relate to the precession of the equinoxes. Can this really be a coincidence, or is the galactic alignment of 2012 actually pointing to an incoming superwave? A large superwave event may well fulfill some of the more apocalyptic prophecies relating to 2012.

LaViolette has an explanation for why superwaves and precession seem to go hand in hand. The sixteenth-century astronomer Nicolaus Copernicus was the first to theorize that the gravitational influences of the sun and the moon on Earth combine to cause Earth to wobble slightly on its axis. This wobble then appears to transcribe a very long circular path in the sky that we call the precession of the equinoxes. LaViolette's alternate view is that galactic superwaves themselves may play the most important role in determining the period of Earth's precession.

He suggests that a superwave might produce a very strong gravity potential wave that could pull Earth toward or away from the galactic center, setting up the 25,771-year cycle. These forces would be greatest when Earth's poles are pointed closest to the direction of the galactic center. This idea would also explain why the periods that LaViolette associates with possible superwave impacts accord so well with those of precession. It would also explain why the conjunction with the galactic center is of such importance.

This would neatly tie up the relationship between the end of the Mayan calendar, the galactic alignment, and their association with catastrophe. However, the traditional theory has not been disproved, and there are some other new ideas about what might be causing precession that also need to be examined. It is also possible that a galactic superwave event may not be as damaging as LaViolette thinks. Small events of a similar type seem to happen about every 500 years and don't tend to trigger major cycles of climatic change. However, the possibility of a more major gravitational wave impact is significant enough that it would be wise not to discount it.

LaViolette's current estimate is that there is a 90 percent chance of a superwave event in the next four centuries. He thinks we should prepare for this and

Are LaViolette's Theories Widely Accepted?

In support of galactic superwave theory, LaViolette points to an impressive fifteen predictions made by the theory that have subsequently been verified. Even so, the scientific community has been slow to accept it. This lack of enthusiasm may be a result of his wide-ranging multidisciplinary approach, especially regarding his interpretation of the Atlantis myths. For mainstream science, any mention of the mythical continent is still very much taboo. Despite this, *Earth Under Fire* remains essential reading for anyone seriously interested in 2012 and the geocosmic cycles that we are all subject to as residents of this planet.

that in the great scheme of galactic evolution, intelligent species evolving on other planets have almost certainly had to face such a challenge. From this perspective, the year 2012 may be a cue for us to reach beyond our solar system in order to find other civilizations that have successfully surfed one of the really big galactic superwaves.

60. Are Plasma Changes Responsible for Climate Change?

The increasing amount of plasma that has been entering our solar system over the last couple of decades has received a lot of attention in the run-up to 2012. A Russian team of scientists, headed by the planet physicist Dr. Alexey Dmitriev, has been following this phenomenon. Their research suggests that this influx of plasma may be responsible for some of the recent dramatic climate changes.

Plasma

Partially ionized gas in which a certain proportion of electrons are free rather than being bound to an atom or molecule. Responds strongly to electromagnetic fields.

The behavior of plasma is quite unlike those of solids, liquids, and gases. In nature, plasmas are usually found in gaslike clouds, as in the case of interstellar nebulae. Other examples of plasmas include ball lightning and the phenomenon of the aurora borealis. A team from the Siberian Russian Academy of Sciences has been investigating changes in the heliosphere, the electromagnetic envelope that surrounds our solar system. The heliosphere acts like a giant protective sheath surrounding our sun and the entire solar system as we travel through space. Normally, it functions as a giant deflector, protecting us from a potentially harmful influx of cosmic radiation and keeping conditions within the inner solar system relatively stable. However, it is now being bombarded with so much radiation that an unprecedented amount is breaking through. This is reaching our sun and all of the planets of the solar system, including our own. The incoming plasma is magnetized to the poles of the earth and concentrates in these regions, creating the effect of the polar auroras. The radiation belts around the planet and Earth's magnetic field are also affected.

The increase in incoming interstellar plasma, Dmitriev suggests, is dramatically impacting the behavior of our sun and its solar system. "Strong evidence exists that these transformations are being caused by highly charged material (in) interstellar space which have broken into the interplanetary area of our solar system," Dmitriev wrote in 1997.

For much of the twentieth century, space was visualized as a near vacuum. The astronomical reality, it is now being discovered, is actually quite different. Our solar system moves through something called the Local Interstellar Space Medium (LISM).

The LISM is not uniformly empty at all, but has greater and lesser amounts of plasmic flux density created by the presence of highly charged particles. The amount of energy within empty interstellar space is actually highly variable. Scientists are now coming to realize that space has more in common with our terrestrial oceans, with their complex tides and currents, than was previously recognized.

The quantity of plasma we encounter in the LISM is a critical variable for what happens in the wider behavior of our solar system. This increased influx of energy is, according to the research of the Russian Academy of Sciences, the fundamental cause of the multiple magnetic and climatic changes that have recently been observed in the sun and across all of the planets. Dmitriev even goes as far as to say that the consequence of the increase in this interstellar plasmic energy is far more important, in his opinion, than human greenhouse gas emissions are in the creation of our planet's current global warming crisis.

61. Further Cosmic Changes in the Heliosphere and the Sun

Both the heliosphere and the sun have exhibited dramatic changes in recent years. The increased plasma flux you just read about has expanded the heliosphere's shock wave in front of the solar system more than tenfold. Dmitriev gives a catalogue of changes he claims this has caused within the solar system:

- Significant physical, chemical, and optical changes observed on Venus; an inversion of dark and light spots detected for the first time and a sharp decrease of sulfur-containing gases in its atmosphere

- The first stages of atmosphere generation on the Moon, where a growing sodium-based atmosphere that reaches 5,500 miles in height has been detected
- Changes in the atmosphere of Mars, including a cloudy growth in the equatorial region and unusual growth in ozone concentration
- Significant melting of the Martian polar ice caps
- A doubling of the magnetic field intensity on Jupiter after the series of impacts from the fragments of the Shoemaker-Levy comet in 1994; also, the appearance of large auroral anomalies, excessive plasma generation, and radiation belt brightening
- Reporting of auroras and a visible increase in brightness on Saturn
- Abrupt large-scale growth of magnetosphere intensity and an increase in brightness on Uranus
- A change in light intensity and light-spot dynamics on Neptune
- A growth of dark spots on Pluto

It seems likely that the increase in this cosmic energy does have some role to play in influencing climate, but it may be one of many contributing factors, rather than a sole cause. Dmitriev himself points out that planetary changes are complex affairs with many interdependent factors. It is the total sum of all these influences that actually determines what happens.

There have also been some recent dramatic changes in the sun. The *Ulysses* spacecraft sent by NASA to measure the magnetic field of the

2012 PERSONALITY:
Richard Carrington

The largest solar flares ever recorded happened on September 1, 1859. This has become known as the Carrington event after Richard Carrington, the young English astronomer who witnessed the event from his private observatory. The flares erupted for less than five minutes. In that time, a huge knot of sunspots appeared and generated a plume that was by far the biggest observed in the 160 years that records have been kept. Before dawn on the following day, a huge firework display of auroral lights bathed Earth, reaching as far south as the Caribbean. The rainbow-hued lights were so brilliant that it was said to be possible to read by them as if it were daylight. The Carrington event also caused major disruption to the telegraph system worldwide.

sun found the magnetic fields of the poles enormously diminished. The magnetic poles of the sun usually reverse at the end of an eleven-year sunspot cycle. At the end of the most recent cycle, the poles only moved to the sun's equator and did not completely invert. This behavior alters everything that was previously believed about the sun's magnetic field. Effectively, the sun no longer has a single north or south magnetic pole; instead, it has four poles located in the equatorial regions.

After the peak of the last eleven-year sunspot cycle in 1999, the sun has had a number of extremely large x-ray flare events. One of these, on April 2, 2001, was so large that it went off the scale completely. The previous scale ran to X-20 as the highest category, but this solar flare had to be categorized as an X-22 event. If the X-22 event had hit Earth, possible consequences could have included major power outages, interruption of the Internet, damage to telecommunications and GPS satellites, and even the wiping of computer hard drives. The most powerful flare observed since then happened on November 4, 2003. It lasted eleven minutes and produced an x-ray flux of at least X-28, although some reports suggest it was much larger even than that.

62. Sunspot Predictions for 2012

Another way to measure cosmic activity is the appearance of sunspots. Astronomer Edward Maunder noted a period from 1645 to 1715 when there were very few sunspots (this time frame is called the Maunder minimum, after him). It was Maunder's studies of this unusual period in history that led to his discovery of the important eleven-year sunspot cycle. At a typical peak of the sunspot cycle, there may be as many as 1,000 spots a year, but during the Maunder minimum the number of spots dropped as low as one or two a year for a thirty-year period. This was also the peak of what has been called the Little Ice Age. This was a period of approximately 400 years, from the fifteenth to the nineteenth centuries, when the drop in temperature was so great that the winter mortality rate in Europe increased dramatically. In London, the river Thames froze over completely every

Sunspots

Magnetic storms on the face of the Sun that form areas of reduced surface temperature.

winter. The edge of the Atlantic ice pack moved southward during the Maunder minimum and glaciers started expanding.

Sunspot activity has been broadly increasing since the Maunder minimum period. This increase has been consistent for more than

Does a Lack of Sunspots Mean Cooler Temperatures Are on the Way?

The sharp downturn in the sunspot cycle may mark the beginning of another minimum period. The result of this would be dramatic cooling, which would impact the agricultural belts of Europe, North America, and Russia, which are responsible for a substantial part of the world's current food supply. For Europe, the possible collapse of the Gulf Stream and its underwater equivalent, the Atlantic warm convector, could signal a massive change in climate. The warming these currents provide prevents European countries from being as cold as those on the equivalent latitudes in North America. Without them, some of the most populated parts of the European continent would be under Arctic conditions.

100 years, but it seems that it reached a peak in sunspot cycle 22 from 1986 to 1996. Sunspot cycle 23 began in 1996 and ended in 2008. The cycle was six months late and weaker than normal.

As of July 2009, the first significant sunspot system has finally emerged. This is the longest period that the sun has been spotless for more than 100 years. Considering the maximum of this cycle is due in 2012, this might seem to suggest that the peak of the cycle may again be less than the recent average. Yet, generally this cycle is still predicted to be 30 to 40 percent more intense than the last one, although NASA has recently downgraded its projections for cycle 24 on the basis of the slow start to the cycle. The official prediction of the NOAA Space Weather Prediction Center is a peak of ninety sunspots in August 2012. There are also some predictions that a delay in the cycle may cause the sun to suddenly burst into violent activity with another series of x-ray megaflares in the X-20+ range, or even cause another Carrington event.

NASA's THEMIS satellite found that a 4,000-mile-thick layer of solar particles has gathered and is growing within the outermost part of the magnetosphere, a protective bubble created by Earth's magnetic field. This is causing a breach in the planet's magnetic defenses. This gap in the magnetosphere is more than four times the size of the Earth itself. This is not a problem at solar minimum, but at peak solar activity it could allow up to twenty times more plasma to impact Earth, making some of the worst solar storms in decades possible.

63. Cosmic Ray Activity in 2012

Recently, scientists have noted increases in cosmic ray activity. Cosmic rays are subatomic particles—mainly protons but also some heavy nuclei—accelerated to almost light speed by distant supernova explosions. The direct effects of increasing cosmic rays include:

- Increased plasma generation in the ionosphere
- Increased magnetic storms in the magnetosphere
- Increased number of cyclones in the atmosphere

The year 2012 is significant in the context of research into cosmic rays because:

- It coincides with the next predicted solar sunspot maximum, and recently discovered breaches in Earth's magnetosphere make us more vulnerable to solar-flare events.

- The effects of increased cosmic dust and radiation entering the solar system will be likely to accelerate by 2012.

According to some scientists, whether the planet cools or heats up depends on the balance of solar activity and cosmic radiation, not greenhouse gases. Dmitriev is one—Henrik Svensmark, the head of Center for Sun-Climate Research at the Danish Space Research Institute, is another. In his book *The Chilling Stars: A New Theory of Climate Change*, he suggests that when cosmic radiation, especially protons, hit Earth's atmosphere, the reaction they cause has the effect of creating clouds. The more cosmic rays there are, the greater the cloud cover.

A shutdown in solar activity and a decrease in the magnetic field of the sun leave our planet more open to the influx of plasmic energy from outside the solar system. This then leads to an increase in cloud cover and the kind of climate change we are now seeing. Svensmark predicts that we could be about to enter a new Maunder minimum–like period and that global temperatures are about to rapidly cool.

Dr. Nir Shaviv, an astrophysicist, also thinks cosmic rays affect our planet's climate. By reconstructing the temperature on Earth over the past 500 million years, Shaviv thinks he has found that changes in the amount of cosmic rays are responsible for more than two-thirds of Earth's temperature changes, making it the most important driver of climate change over long periods of time.

Shaviv hypothesizes that the sun's passage through the spiral arms of the Milky Way appears to have been the cause behind the major Ice Ages over the past billion years. He has correlated variations in the cosmic-ray flux to the solar system's orbit around the center of the galaxy and through its spiral arms. In the more crowded spiral arms, such as the Orion arm, which our Sun is currently in, there is a higher density of cosmic rays. Shaviv agrees with Svensmark that the result of this increase is that Earth becomes cooler.

Both Svensmark and Shaviv are considered "climate skeptics" who dispute the extent to which the creation of greenhouse gases is contributing to the current climate change.

64. What Does Increased Cosmic Activity Mean for Earth?

In the approach to 2012, Earth is undergoing a variety of major geophysical changes unprecedented in scale and effect. The causes of some of these appear to be man-made, but others appear to be responses to changes in the behavior of the sun and an increase in cosmic radiation. There is

Solar Activity Definitely Affects Our Upper Atmosphere

Svensmark points out that it is actually well established and uncontroversial that solar activity has a direct influence on the eleven-year variation of stratospheric pressure levels found in the upper atmosphere. The electromagnetic fields of our planet are highly sensitive and respond to a range of influences from solar wind to tropical storms. A tenfold increase in cosmic radiation is likely to affect these fields and the upper and lower atmosphere of the planet in ways that may be unpredictable. The debate is to what extent and how these changes will manifest.

widespread speculation linking these events with a planetary catastrophe in 2012, possibly a reversal of Earth's poles.

Other important changes are also happening. There appears to be an increase in the number and severity of extreme weather events such as tropical storms, flash flooding, droughts, and tornadoes. Major geophysical events such as earthquakes and volcanic activity appear to be increasing substantially. For instance:

- In the last fifty years, the number of tornadoes has doubled and there has been a dramatic increase in the geographical area in which tornadoes are found.
- According to Michael Mandeville of *www.earthchanges-bulletin .com*, the number of recorded earthquakes in the last century has increased by 500 percent and measurements of volcanic activity are up by at least 200 percent. This may be at least partly due to the increase in the number of seismic and volcanic detectors around the world. The spread of humans into what were once remote parts of the world may also be a factor.

The fact that Earth's climate is in a state of rapid change is now well established. Climate change is a notoriously difficult area in which to make accurate predictions. One of the major problems is that science has become very specialized, which makes assessing all the different factors that act on a system as large as a planet challenging. While the unusually high concentration of carbon dioxide in the atmosphere is certainly having some effect on climate, the influence of solar activity and cosmic rays on global temperature may be equally, if not more, important. If the sun's behavior changes, Earth's climate is likely to respond. Assessing these trends and the balance of probabilities is a matter of fierce debate.

65. The Magnetic Poles of Earth Are Shifting

Dmitriev's point of view on the pole shift is that it is already happening. In fact, he believes that the shift actually began in 1885. In the last 100 years, Earth's magnetic south pole has traveled almost 560 miles toward, and into, the Indian Ocean. The magnetic north pole moved more than 170 miles between 1973 and 1994 in the direction of Siberia via the Arctic Ocean. The rate of the magnetic pole's movement has also increased in the last century compared with fairly steady movement in the previous four centuries.

Oregon State University researchers investigating the sediment record from Arctic lakes have been able to use carbon dating to track changes in the magnetic field. They found that the north magnetic pole has shifted significantly in the last 1,000 years. It generally migrated between northern Canada and Siberia, but has occasionally moved in other directions. The causes of these magnetic changes are related to changes in behavior of the electrical flow in the iron at the core of the planet. This, in turn, is influenced by incoming plasma at the poles of Earth.

Earth's magnetic field is not uniform and is becoming less so. There are a number of areas called "world magnetic anomalies" that

How Should We Prepare for a Pole Shift?

Dmitriev estimates the speed of the pole shifting process will increase to around 125 miles or more a year in the near future, and that we should prepare for the consequences of this in a globally coordinated way. The appropriate response, he says, should be to draw up a "global, ecology-oriented, climate map which might reveal (the location of) these global catastrophes."

generate a substantial magnetic field independently of the two poles. The four most significant ones are in Canada, Siberia, Antarctica, and Brazil. These anomalies have recently undergone significant growth. Dmitriev thinks the movement in the magnetic poles and the growth in magnetic anomalies indicate that something very dramatic is going on in the core of our planet. The scale of these changes indicates something beyond even the magnitude of the Gothenburg magnetic flip event that happened around 12,700 years ago, when the magnetic poles migrated to near the equator. He believes the signs suggest a complete magnetic pole reversal is already underway.

66. The Possibility of a Physical Pole Shift

Patrick Geryl, author of *How to Survive 2012*, strongly believes a complete magnetic pole reversal will inevitably trigger a disastrous physical pole shift, simply because Earth's core is iron and therefore will respond to the new polarity. This is an overly simplistic view that is not supported by scientific or historical evidence.

There is no evidence that a physical pole shift happened during previous magnetic pole shifts and there is nothing to indicate why it should happen this time. An event like this happened once before in the geological record, but not for many hundreds of millions of years. Geryl's belief that a pole shift is certain in 2012 has led him to conclude that the only reasonably safe places to be in such an event will be in special unsinkable ships or deep underground, high up in a major mountain range. Even then, survival is not guaranteed.

Geryl believes Earth reversing its direction of spin will initiate the pole shift. This idea comes from Gregg Braden's book *Awakening to Zero Point*, in which he examines a scenario where Earth's rotation actually slows,

Physical pole shift

A situation in which the planet actually rolls over on its axis. A physical pole shift would likely be catastrophic for the global ecology. One probable consequence would be major crustal displacement, as the flip causes tectonic plates and continents to collide with each other.

momentarily stops, and then reverses in the opposite direction. This theory would require an unknown force to negate Earth's spin, slow it to a halt without tearing the crust of Earth from its mantle, and then reverse the force so that Earth spins in the opposite direction. The forces responsible for the minuscule degree of slowing that the Earth is already undergoing would in no way be adequate to do this, nor would any other known force in our solar system.

On the other hand, a good indicator of the possibility of changes in the physical poles of Earth is an effect called the Chandler wobble. This is the change in the spin of the earth on its axis. It's named after Seth Carlo Chandler, an American astronomer who first discovered the wobble back in 1891 after thirty years of observations. The effect causes Earth's physical poles to move in an irregular circle. This wobble has a seven-year cycle. The wobble:

- Produces a very small ocean tide, the pole tide, which is the only tide not caused by bodies outside Earth.
- Has varied in amplitude since its discovery, reaching its largest size in 1910 and fluctuating noticeably from one decade to another.
- Is caused by fluctuating pressure on the bottom of the ocean, caused by temperature and salinity changes and wind-driven changes in the circulation of the oceans, according to the Jet Propulsion Laboratory.

67. Technology's Impact on the Biosphere

One of the variables in the process of rapid magnetic change that Dmitriev reports is the effect our industrial and technological civilization is having on our planet. More than 30 percent of disturbances in the magnetosphere are now caused by electricity production, transmission, or consumption. For example, consider the Van Allen radiation belts over the eastern United States. These are two belts of plasma surrounding Earth that are held in place by the planet's magnetic field. The inner belt extends 200 to 6,000 miles from Earth's surface and has a high concentration of protons. The outer belt extends 12,000 to 26,000 miles and is made of electrons. These belts have moved inward from more than 200 miles above the surface of the planet to slightly more than six miles. The Van Allen radiation belt is caused by the massive amount of energy being transmitted between the power stations around the Great Lakes to the eastern seaboard.

Transmission between power stations is just one of the many ways we are changing our electromagnetic environment, potentially with unforeseen consequences. For instance, some species and habitats are more sensitive than others to the effects of these changes. The rapid decline in bee population,

2012 PERSONALITY: Michael Mandeville

Independent researcher Michael Mandeville of *www. earthchanges-bulletin. com* has been exhaustively analyzing trends in seismic and volcanic activity from around the world. Using a very detailed statistical analysis, Mandeville claims to have found correlations between the position and motion of the pole with increases and decreases in earthquakes and volcanic eruptions. These correlations are sufficiently consistent, he claims, to conclude that the Chandler wobble stresses Earth's crust, which in turn creates a cycle of earthquakes and volcanic activity.

known as Colony Collapse Disorder (CCD), may be a symptom of the change in the earth's magnetic polarity. A survey commissioned by the Apiary Inspectors of America found losses of more than 30 percent in the bee population from CCD. Some scientists estimate that life on Earth is so dependent on bee pollination that the current human population would be unable to feed itself just eight years after the collapse of the bee colonies.

While some scientists believe that bees find their hives by following polarized lines of light in the sky, research at National Tsing Hua University of Taiwan into magnetic reception in bees has shown the presence of magnetite, which suggests they have magnetic homing senses. Thus, changes in Earth's magnetic field and the influence of man-made electromagnetic pollution are possible causes of the dramatic bee decline.

Magnetosphere

A highly magnetized region around Earth that protects the planet from incoming cosmic radiation.

Whales may also have a biomagnetic sense, which allows them to navigate by sensing Earth's magnetic fields. Whales following magnetic field lines could accidentally beach themselves in areas where the field lines intersect with the coast. A study in the United Kingdom by biologist Margaret Klinowska found a correlation between local magnetic field lines and sites where whales were stranded on shore. The biomagnetic theory may also explain why there are multiple-species strandings. The use of underwater sonar has also been implicated in whale beachings.

The earth's weakened magnetosphere allows more ultraviolet light to penetrate through the atmosphere to the surface. Frogs and other moist-skinned amphibians are among the species most sensitive to these increases in ultraviolet radiation. There has been a recent sharp decline in frog and other amphibian populations in both tropical and temperate climates.

68. Responding to the Electromagnetic Crisis

If the rapid increase in man-made electromagnetic emissions is left unchecked, there will likely be consequences for human health and the health of the biosphere. The combination of a number of other geophysical effects converging in 2012 may mean that this impact is compounded. These effects include:

- Weakened magnetosphere
- Solar maximum predicted for 2012
- Increase in interstellar plasma and cosmic rays
- Movement of the magnetic poles

One possible scenario is that at the solar maximum around 2012, a massive solar eruption on the scale of the Carrington event (see Part Three) could pass through the weakened magnetosphere of Earth. This could massively impact our global communications systems and computer networks and dramatically accelerate the changing motion of the magnetic poles. If the magnetic flux

What We Could Do to Alleviate the Effects of a Large Solar Event

If a really large solar event occurred, not only might our electromagnetic infrastructure be damaged, but the resulting impact on the biosphere may require us to act to stabilize the electromagnetic field of the planet. This may require turning off sources of electromagnetic disruption that are currently considered essential services. These include electricity power grids, mobile phone networks, satellite communications networks, ELF communication systems, radar, and microwave networks like WiFi and CCTV monitoring systems.

of the solar flare event is of sufficient magnitude to overwhelm the ring main of Earth's magnetic field, it could theoretically produce a rapid magnetic pole shift.

The sun's recent behavior does suggest that major solar eruptions are likely at the next solar maximum. The Carrington event megaflare happened at the end of the 300-year-long solar shutdown of the Maunder minimum period. This was followed by more than 100 years of increased solar activity on the sun. During this period, the strength of the sun's magnetic field more than doubled. The recent decline in the sun's polar magnetic field may mark the end of that warm period.

It may be that during the shift to a colder period, the sun's behavior goes into oscillation between less and much greater activity, increasing the likelihood of megaflare events. It is also possible that a Carrington event megaflare could signal the beginning, as well as the previous ending, of one of these periods of much decreased solar activity and colder temperatures on Earth. The flare that caused the shutdown of the Canadian power grid in 1989 was rated

Solar maximum
Time period of maximum sunspot counts.

as an X-20 event; a Carrington event–sized flare could be more than twenty times that size. Additionally, the substantial hole in the Earth's magnetosphere detected by NASA's THEMIS probe means that our planet is much more vulnerable to incoming solar radiation than was previously realized.

PART FOUR

Outer Space's Impact on the Earth and 2012

From plasma and orbs to the Oort cloud and Planet X, there's a lot going on in outer space that figures into the discussion of 2012. Chances are you've never heard of some of these phenomena.

In this section you'll learn all about the otherworldly happenings that may or may not cause some dramatic events here on Earth. Read on, and be amazed.

69. The Relationship Between Atmospheric Plasma and Tornadoes

The mechanics of tornado formation are not well understood. Spectators close to tornadoes, which are also sometimes called funnel clouds, have reported little or no wind surrounding the funnel cloud while it is airborne. But when those same tornadoes touch the ground, they can have devastating effects, easily destroying houses, lifting cars into the air, and displacing objects and people—in some cases, for miles.

The role that plasma plays in the behavior of our planet's climate is remarkable enough, but scientists and researchers have shown that plasma vortexes can be found in tornadoes and in many unusual atmospheric phenomena that appear to be increasing in frequency. Recently declassified government documents have shown that many UFO sightings may also be related to atmospheric plasmas.

There are many names given to the glowing plasmas that are found in the atmosphere, and there is a broad spectrum of these self-luminous effects that includes ball lightning, upper atmospheric effects called sprites and elves, earthquake lights, the aurora borealis or northern lights, and glowing orb-like balls of light.

Dmitriev calls all of these natural self-luminous formations (NSLFs), or vacuum domains. For simplicity, they are referred to here as plasmoids, which is defined simply as any coherent structure of plasma and magnetic fields. Plasmoids have been proposed to explain natural phenomena as diverse as ball lightning, magnetic bubbles in the magnetosphere, objects in cometary tails, and structures found in the solar atmosphere.

Plasma physicist Dr. T. Matsumoto has shown that atmospheric plasmoids exhibit various strange behaviors, such as hopping over land, skimming across water surfaces, and passing unchanged through glass, water, and air. Changes in atmospheric pressure and other conventional meteorological explanations are unable to explain these anomalies. Self-luminous formations or plasmoids, as proposed in the Dmitriev model, are luminous effects created by the combined forces of electromagnetism and gravity because of these cosmic collisions. It is an elegant theory, but a hugely controversial one for conventional meteorology. The incidence of these formations, Dmitriev believes, is related to solar activity. Dmitriev believes that similar storms found on other planets are created by the same mechanism and increase in frequency when the sun is most active. Essentially, tornadoes on Earth can be seen in this model as a sympathetic response to similar storms on the sun.

70. Are Some Tornadoes Caused by Mini Black Holes?

Many viewers also report seeing lights at the center of tornadoes. Dmitriev's department of the Russian Academy of Sciences has collected reports of a wide variety of luminous eye phenomena in tornadoes that have been witnessed by observers.

These include:

> "[a] Ball of fire . . . lightning in a funnel . . . yellow shining surface of [a] funnel . . . incessant lightning . . . [a] fiery column . . . glowing clouds . . . [a] brilliant luminous cloud in a funnel . . . beaded lightning . . . exploding fireballs . . . [and] a rotating band of deep blue light."

The conclusion of their research has been that these plasmoids are inherent to tornadoes and that it is the vortexes created by these plasmoids that are responsible for creating the funnel clouds, not the other way around. This differs from the conventional meteorological explanation, known as the Brooks model. The Brooks model states that tornadoes are a thermodynamic effect produced by a parent storm cloud. However, Dmitriev cites examples of tornadoes appearing without any clouds overhead. In cases like these, what appeared to be a parent cloud actually formed around the central vortex as the tornado progressed. In Dmitriev's model, the tornado is actually *caused* by the rotating central tube of plasma. This model has the advantage of being able to explain the anomalous behaviors that occur in and around tornadoes that the conventional Brooks model cannot.

Effects of tornadoes that are better described by the plasma vortex model include:

- Funnel clouds are seen to travel over the surface of the earth in jumping movements, suggesting that other forces besides atmospheric pressure are causing them.

- When a tornado is just slightly above the ground, no lifting effects are seen, but as soon as it touches the ground levitation begins. The area inside the tornado has been measured to have a lower air pressure, but this does not occur beneath the cloud.
- As tornadoes cross rivers, they have been seen to form trenches in the water, sometimes up to twenty feet deep. This suggests that they are sources of powerful, anomalous gravitation.
- Tornadoes have been observed to emit pitches or hissing noises while airborne, indicating very high levels of electrostatic charge.
- Tornadoes can transport objects and even living creatures over long distances without damaging them. This suggests that the lifting power of a tornado is being caused by levitation, not by suction and rotation.
- Many types of luminous phenomena have been associated with tornadoes. Lights have been seen both before a tornado appears and inside its central vortex.

Dmitriev has attempted an alternative explanation of how tornadoes and their funnel clouds arise. He suggests they may originate in small atmospheric holes made in the upper ionosphere by tiny microcomets. These holes are well documented and have been seen and recorded in the ultraviolet spectrum. The cause of these holes is thought to be incoming helium nuclei from the sun that react with the upper ionosphere to create miniature black holes. These mini black holes of spinning gravitational waves create a vacuum effect and a pulsed heat release. This, in turn, produces a wide variety of self-luminous phenomena, depending on a number of variables.

71. Increased Plasmic Weather Phenomena

Tornadoes are just one example of plasmoid activity in the atmosphere, which also creates a wide range of other weather effects. The incidence of these phenomena seems to be increasing significantly in frequency and scale. For example, sprites, elves, and blue jets are all transient luminous event (TLE) atmospheric phenomena that are observed at high altitudes above thunderstorms. Passengers aboard planes have recorded some of the best footage of sprites. These plasma effects can be both huge in scale and spectacular in intensity.

Sprites are usually reddish in color and are among the oldest known of these phenomena, first reported more than 100 years ago. Sprites appear at heights of between thirty and fifty-five miles up in the atmosphere as glowing plasma-conducting columns up to several miles in length. Sprites typically last milliseconds, just long enough to easily be perceived by the human eye. They are often carrot shaped and scatter radio frequencies.

Elves are a related type of discharge found at around fifty-five miles up in the atmosphere and above that are described as an enhanced airglow in an expanding doughnut-shaped ring. Elves result from an especially powerful electromagnetic radiation pulse (EMP) that emanates from lightning discharges. Elves last for less than a thousandth of a second, which makes them virtually impossible to see with the naked eye.

Blue jets are a strange new type of effect that typically consists of an upwardly moving, bluish beam emission from cloud tops, known as jets. These occur below thirty miles in altitude and last up to a quarter of a second.

Other effects associated with these three phenomena include gravity waves that generate upper ELF signals of around 300kHz and horizontal magnetic field variations in the frequency range of the earth-ionosphere cavity resonance.

Dr. Massimo Teodorani and his associates have extensively researched anomalous plasma formations that have been seen in the valley of Hessdalen in Norway. For more than two decades, many eyewitnesses in the valley have observed and reported flickering, pulsing lights that change shape. Teodorani's team of Italian astrophysicists joined a team of Norwegian engineers lead by

Prof. Erling Strand in a joint study using radar, photography, video, radio spectrum analyzers, and spectroscopes.

The phenomenon observed by the team of scientists could be broken down into two groups: 95 percent were thermal plasmas and 5 percent were unidentified solid objects. The plasmas emitted long-wave radio frequencies and appeared when there were disturbances in the geomagnetic field.

The summary of their research makes the following conclusions:

1. Most of the luminous phenomena are thermal plasma.
2. The light balls are not single objects but are constituted of many small components that are vibrating around a common center.
3. The light balls are able to eject smaller light balls.
4. The light balls change shape all the time.
5. The luminosity increase of the light balls is due to the increase of the radiating area. However, the cause and the physical mechanism with which radiation is emitted is currently unknown.

72. Project Condign and UFOs

Project Condign was the name given to a top-secret UFO study undertaken by the British government's Defense Intelligence Staff (DIS) between 1997 and 2000. This was declassified and released to the public in 2005 after a request under the Freedom of Information act by the UFO researchers Dr. David Clarke and Gary Anthony.

The report looked into all reported sightings of UFOs over the previous three decades. Between 1967 and 1997, there were 100 to 750 credible incidences a year in British air space alone. The project came to the conclusion that the phenomenon was indisputably real, but that they weren't dealing with physical craft. Instead, the report concludes that the UFO phenomenon is caused by highly charged atmospheric plasmas. These were named unidentified aerial phenomena (UAPs).

In a high-energy state, UAPs are visible to the naked eye, but in a lower-energy state they may only be visible in the infrared spectrum or on radar. The UAPs generally have a magnetic field about 5,000 times weaker than Earth's, but the UAPs field is pulsed and rotating, which seems to provide them with their source of propulsion.

2012 PERSONALITY: Susan Joy Rennison

Plasma effects in the atmosphere are not confined to the small scale. They also appear as very large spheres that have been reported by Dmitriev. These are also described by Susan Joy Rennison in her book *Tuning the Diamonds*, which makes a detailed argument for a relationship between the increase in these new electromagnetic phenomena of all scales and a shift in human spiritual evolution. These spheres vary in size from around thirty feet across up to two miles in diameter. In one case, on a data-gathering expedition in the Altai mountain range of Russia, scientists from Dmitriev's team measured a large-scale formation that was nearly five miles wide.

The Effect of UAPs on Humans

Proximity to UAPs has affected people in various ways. When questioned, 75 percent of people said they felt odd or dizzy or reported tingling sensations, and 65 percent saw vivid images and/or said they experienced pleasant vibrations. Forty percent said they experienced fear or terror. The Condign Report also notes that humans who encountered the UAPs exhibited serious perceptual and temporal distortions. The Condign Report's conclusion after three years of gathering and processing data is that there is "no evidence that they (UAPs) . . . represent any hostile intent" and that "further investigation should be made into the applicability of various characteristics of plasmas in novel military applications."

The UAPs were also capable of flying in formation and frequently formed complex geometric arrangements of three to five balls, reminiscent of the underside shape of a classic UFO. These balls were able to move synchronously and were observed to be able to join and then part again. When flying in formation, the balls of plasmas often give the appearance of a solid craft.

The power of these UAPs is so great that their plasma fields can seriously harm a nearby vehicle or person. The Project Condign report notes that scientists in the former Soviet Union have taken a particular interest in the UFO/UAP phenomena and are pursuing related techniques for military purposes. Several Russian and Chinese aircraft have been destroyed chasing UFOs.

73. Are Plasmas Alive?

Extensive research on these plasmoids has also been carried out by the Russian Academy of Sciences in a department headed by V. G. Azhazha. They concluded that these phenomena are highly mobile high-energy plasma vortexes, which they call UFOs. Their studies showed that these plasma UFOs exhibited the following behaviors:

- Gradual growth
- Splitting into two or more separate parts
- Dissolution to invisibility
- Disparate bright lights merging into larger formations (often reported as small craft joining the mother ship and forming a row of portholes)
- Disappearance, accompanied by a smell
- Rotation, nonlinear motion
- Weak thermal radiation
- Translucence, haloes, blackness
- Beamed light emissions, described by the Russian scientists as "discharge, or leakage paths" that allow the plasma to "float off" and to change direction, seemingly at will.

Given the properties of these plasmas, scientists have been debating whether they could be considered a form of life. The physicist David Bohm, who was one of the primary research partners of Albert Einstein, remarked that he frequently had the impression that plasmas were alive and that they had many of the properties of organic life. Indeed, plasmas can imitate the functions of a simple cell. For instance, they have a semipermeable cell wall through

Earth Is Witnessing Increased Plasmic Phenomena

The number of plasmic and plasmoid phenomena of all kinds observed on our planet appears to be increasing as we approach 2012. This could be directly related to the increase in interstellar plasma that our solar system is encountering. As this combines with the next solar maximum due for 2012, the amount of plasmic effects in Earth's atmosphere are likely to increase even further. Plasma physicist Matsumoto has shown that their frequency is correlated with increases in solar activity. The range of their effects varies from small orbs that are barely visible all the way up to giant atmospheric effects more than a mile in diameter. The aurora borealis is a plasmic effect on a planetary scale and is even more dramatic seen in the ultraviolet part of the spectrum.

which they can "feed" by absorbing other less-organized plasma. In fact, plasmas were named for their similarities to living blood cells.

V. N. Tsytovich, another scientist based at the Russian Academy of Science, has shown how plasmas can self-organize when exposed to electrical charge. Tsytovich has developed his observations into a theory of what he calls inorganic life. The principle location of inorganic life is in the helical dust structures that have been seen to form around stars and in interstellar space.

In a gravity-free environment, these plasma particles bead together to form string-like filaments, which then twist into helix-shaped strands closely resembling DNA. These structures are electrically charged and are attracted to each other. They "feed" by assimilating other less-organized plasmas through their boundary walls. They "reproduce" by amoeba-like splitting, and each of the plasma's offspring can also grow, feed, and reproduce.

Mircea Sanduloviciu of Cuza University, Romania, has also worked extensively on plasmas. Sanduloviciu has developed a theory that plasma spheres arising within electric storms were the first cells on Earth. Sanduloviciu believes that the emergence of these plasma spheres is a prerequisite for the evolution of biological cells and that cell-like self-organization can occur in a few microseconds. Like the Tsytovich helical dust structures, Sanduloviciu found that the spheres could replicate by splitting into two.

74. Orbs Are Showing Up in Digital Photos

Some researchers believe that plasma may be responsible for the phenomenon of orbs, which are glowing balls of light that are increasingly appearing in digital photography. But before they started showing up in photos, orbs appeared in the folklore of many cultures. The floating ghost lights often inspired awe and fear in those who encountered them. They usually hovered close to the ground or between the trees. They were considered both intelligent and powerful and were to be treated with respect. Folk names given to floating orbs include corpse light, fair maids, merry fires, foxfire, friar's lantern, hinkypunk, hobby lantern, ghost light, jack-o'-lantern, kitty-with-a-wick, peg-a-lantern, pixy-light, spunkie, and walking fire.

Orbs are usually invisible to the naked eye but appear, sometimes in abundance, when photos are taken. There are several possible explanations for this. First, more pictures are being taken now because digital photography makes it so easy. Second, the optics of digital cameras are more suited to capturing orbs than those of conventional film. Third, a percentage of orb phenomena are caused by simple physical reasons:

- Solid orbs: dry particulate matter such as dust, pollen, etc.
- Liquid orbs: droplets of liquid, usually water, e.g. rain
- Foreign material on or within the camera lens and body

These factors do not quite account for some of the more unusual properties of these orbs, nor do they explain why orbs tend to appear under certain circumstances and not others. In addition, many of the orbs that appear in digital photos show complex internal structures, with patterns much like those of snowflakes. In addition, they often have cell-like membrane structures around their edges. Orbs appear in many millions of digital pictures and it has been estimated that as many as one-third of all digital photos contain anomalies that are orb-like in appearance.

The number and density of orbs in photography is correlated with a number of certain conditions. A greater number appear:

- At the beginning of lightning storm activity, which tapers off once the frontal boundary passes.
- When a low-pressure zone passes over or barometric pressure is dropping.
- When a Tesla coil device is being operated.

- Around high-power electricity cables and towers at night.
- In crop circle formations.
- At gatherings of people focusing intentional energy, including ceremonies and meditations.

75. Orbs' Possible Relation to Crop Circles

The appearance of orbs seems to relate not only to atmospheric and electrical conditions, but also to human intention. Researchers Kris and Ed Sherwood have conducted guided meditations specifically designed to connect with the energies of the orbs. They have conducted these light-ball visualizations at sacred sites and in crop circle formations around the English countryside.

The Sherwoods have been remarkably consistent in sighting orbs. Ed has taken digital photographs while Kris has been meditating, and the resulting photos sometimes show hundreds of orbs, many of them bright and well defined.

2012 PERSONALITY: David M. Rountree

David M. Rountree, AES, has been studying the phenomenon of orbs closely and has named them Unified Field Plasmoids. He developed a theory that the orbs he was capturing in his photographs were composed of plasma. He hypothesized that a camera flash could render them photographic. Around 1990, he saw an orb without the help of photographic equipment. He reported it was three-dimensional and slightly fluorescent, which led him to theorize that some orbs could attract electrons to make themselves more visible. These electrons, by their very nature, needed to discharge themselves, so they headed for the ground. Once an orb hit the ground, it would "effectively disappear in thin air from whence it came," Rountree reported.

In some cases, motion blurs suggest the orbs are moving faster than the shutter speed of the camera. In other pictures, luminous objects appear to be hovering above peoples' heads. The results were particularly noticeable inside crop circle formations.

The majority of crop circles recorded in the last twenty years have appeared in the west of England, but the earliest recorded image resembling a crop circle is depicted in a seventeenth-century English woodcut and pamphlet called the Mowing-Devil. The image depicts the devil with a scythe cutting a circular design in a field. The accompanying text states that the farmer, disgusted at the wage his mower was demanding for his work, insisted that he would rather have the devil himself perform the task.

Crop circles

Also known as crop formations; patterns created by the flattening of cereal crops such as wheat, barley, rapeseed, rye, and corn. These often take the form of intricate geometric patterns.

Human-created crop formations have been documented by a number of different groups of circle makers since 1978. This has been interpreted by some to mean that all crop formations are hoaxes, deliberately created to fool people into thinking they are some kind of supernatural phenomenon. However, this ignores the fact that various data readings taken in crop formations of unknown origin have consistently shown marked differences from readings taken in crop formations of known human origin.

It also does not take into account the reports of many human circle makers, allegedly responsible for the "hoax" crop formations, who say they have often witnessed anomalous lights and floating luminous orbs in fields they were working in. This has led to the speculation that crop formations may be an interactive phenomenon and that human circle makers can attract the attention of the genuine circle makers with their activity. It may be possible that a form of communication between the symbology of art and circle-making activities may be occurring in these cases.

76. Do Plasma-Vortexes Cause Simple Crop Circles?

The first scientific attempt to explain the appearance of crop circle formations was made by Terence Meaden. Meaden is a physicist and meteorologist who is also the founder of the tornado study group TORRO, which collects, analyzes, and publishes data on the incidence, strengths, and origins of tornadoes and severe storms.

TORRO also studies the ball lightning and other phenomena associated with these events. One of his major areas of research has been the study of tornadoes and plasmoids. Being very familiar with properties of atmospheric plasma, Meaden hypothesized that a whirlwind or tornado could potentially generate a plasma-vortex that could account for the delicate and precise way that crop stalks appeared to be laid down in circles, as if from above. The plasma-vortex theory could also explain the anomalous lights that were associated with crop circles as friction-generated plasmoids created by a plasma vortex.

A professor in Japan was able to produce laboratory results that seemed to corroborate Meaden's whirlwind-created plasma-vortex theory. Using electrostatic discharges and microwave interference, a vortex modeled on interactions between a spinning electrical field and Earth's magnetic field was simulated, showing that it was theoretically plausible.

Meaden's theory was a promising way to explain the phenomena of the more simple circles that appeared in the 1980s. However, as much more complex formations started to appear, the idea that these could credibly be created by a simple whirlwind was dispelled.

The plasma-vortex theory was still embedded in the traditional meteorological notion that whirlwinds generate plasmas, rather than the other way around. There was simply no way to argue that more complex crop circles were being created merely by a combination of wind and electrical charge. Meaden had always contended that the causes were more complex, but the whirlwind-based plasma-vortex theory had hit a major roadblock.

77. Grain Samples from Crop Circles Can Provide Interesting Data

The plasma vortex theory took a back seat until Dr. William Levengood redeveloped it. Levengood attempted to explain the microwave energy and other residues he had found in grain samples taken from crop formations. In fourteen years of scientific investigation into crop circles, Levengood has examined more than 250 crop formations in detail. Since 1992, he and his team have been involved in extensive on-the-ground surveillance and the collecting of samples and electromagnetic and other readings from crop formations. These samples are then sent to Levengood for analysis in his private laboratory based in Michigan, where he compares the grain collected from in and around the circles with control samples taken from edges of the fields that the formations were found in.

One notable characteristic of grain samples taken from the formations was the incidence of exploded wheat nodes. Wheat stems taken from crop formations revealed a massive number of expulsion cavities (holes blown out at one or several of the nodes). This effect suggests that they were subjected to an internal pressure so sudden and powerful that it blew holes through the node points of the stem walls as the internal sap rapidly expanded. Levengood concluded that this could be characteristic of electromagnetic radiation, probably microwave, emanating from the epicenter of the formation.

Node

The joint on a stem where a leaf is attached.

Typical crop samples taken from formations also include the following characteristics:

- Stalks are very often bent up to ninety degrees without being broken, particularly at the nodes.
- Stalks are unusually enlarged, stretched from the inside out by something that seems to heat the nodes from the inside. Sometimes this effect is powerful enough to literally explode the node, blowing holes in the node walls and causing sap to leak from the stalk.

- Stalks are left with a surface electric charge, suggesting the force that flattened the crops was electrical.
- The thin tissue surrounding the wheat seed shows an increase in electrical conductivity, consistent with exposure to an electrical charge.

The germination of some seeds found in crop formations has accelerated growth and vigor. Yet there are also significant reductions in the growth rates of seedlings germinated from the wheat seeds taken from formations; seedling heights are 35 percent shorter than controls. These results, Levengood concludes, strongly suggest rapid heating, such as would be caused by the exposure of the plants to microwave radiation or unusual electrical fields.

Levengood's laboratory analysis of samples from crop formations also notes the presence of a gray dust of hematite or iron oxide, which is of a type found in meteoritic material. This suggests to Levengood that the formations were created by a close encounter between a meteoritic system and a plasma vortex. The vortexes described in Levengood's theory are created much higher in the atmosphere than those proposed by Meaden's theory. Beginning in the ionosphere, where there is an abundance of microscopic meteoritic material, the Levengood vortexes are drawn down to the surface of the earth, apparently to areas of significant electromagnetic charge. This could be related to underground geomagnetic fields in the locations where the circles are found. Measurements taken by Levengood and his team in recently made formations have shown elevated magnetic levels that quickly dissipate, suggesting that a charge has somehow been released.

78. Unexplained Fractal Crop Formations

The next significant level of evolution in crop formation design occurred in June 1996 when a formation appeared near Alton Barnes in Wiltshire, England. It consisted of a series of eighty-nine circles arranged in a pattern that seemed to strongly suggest the double-helix structure of the DNA molecule. A large number of circles of changing scale were laid out with a remarkable degree of exactitude in the formation. Videographer Peter Sorensen was living just a few hundred feet away from the field in which it appeared; as the first person to discover it, he seized the opportunity to film the formation at dawn before anyone else had a chance to enter. His footage shows the crop formation to be executed very precisely with tightly woven circles with well-defined edges.

The Progression of Crop Circle Designs

Many of the initial crop formations in the 1980s were simple circles and consisted of singular or multiple circles, sometimes arranged in geometric patterns. The first major evolution in the complexity of crop formations happened in 1990 with the first pictograms. These were often comprised of a series of geometric shapes arranged around a central shaft or bar. A famous example of one of these pictographic formations can be found on the cover of the Led Zeppelin box set that appeared in 1990. A similar type of design was created for the movie *Signs*.

▲ The DNA double helix crop circle formation

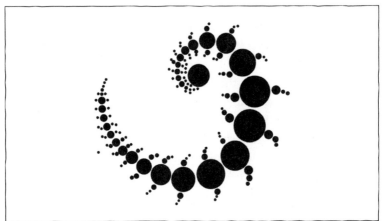

▲ The Julia-set crop circle formation

One of the most enduring mysteries associated with crop formations is related to the appearance of a large crop circle within sight of Stonehenge in broad daylight. The formation was in the bass clef–like shape of a Julia-set fractal that appeared in full view of the busy A303 road within a thirty-minute period one Sunday afternoon in July 1996.

Tourists spotted the 900-foot formation just after 6 P.M., and a pilot, a gamekeeper, and a security guard confirmed it had not been there half an hour before. The complex curving arc, precisely composed of many circles of changing scale, was certainly a relative of the double helix DNA formation that had appeared earlier that summer near Alton Barnes. The undetected, near instantaneous appearance of such a large formation in an area with so many potential observers has defied any attempted explanation.

Three weeks later, at the end of July, another even more spectacular Julia-set fractal appeared at the Windmill Hill megalithic site, near Avebury stone circle. This time the formation was larger, had three spiral arms, and contained 194 circles! The geometric relationship between the two was very clear.

The symbolic language of crop circles seems to evolve by establishing a motif, then expanding upon it with further formations that become increasingly elaborate. Skeptics have suggested that this can be accounted for by the fact that hoaxers are getting better and improve throughout each season. But no hoaxers have ever claimed to have made either of these stunning formations. Much smaller man-made designs have taken many hours to complete, often with elaborate equipment. These man-made formations often include telltale signs of crop damage and tracks left by their makers into and out of the formation.

79. Might Crop Formations Contain Astronomical Messages?

The study of crop formations has attracted geometers and mathematicians eager to interpret the shapes and measurements of the formations. The most fruitful of these endeavors has been the interpretation of astronomical messages in the crop circles, mostly related to cycles of the sun, moon, and other planets. Some articulate the geometric relationships inherent in these cycles. Others seem to explicitly contain messages.

One of the message crop formations is the so-called missing-Earth formation that appeared at Longwood Warren, Wiltshire, in 1995. An orbital ring containing sixty-five circles surrounded a series of concentric circles or orbits, complete with circles for planets. This was widely interpreted as being a map of our inner solar system on a certain date. It seemed to show the orbits of the asteroid belt (the orbital ring), the orbits of Mars and Venus, and the position of Mercury with the Sun clearly marked at the center. The proportions of the orbits were remarkably accurate. The one exception was that although Earth's orbit was shown in the formation, Earth's position was not!

This led many interpreters to suggest that the formation was foretelling a catastrophic end to our planet on a specific date. One possible date corresponding to the alignment shown was on the inferior conjunction of Venus on January 16, 1998. The lineup also occurred again in June 2004, during another Venus loop (the sixty-five day period where the planet goes retrograde and appears to move backward in the sky as observed from Earth) and four days before the Venus transit. No apocalypse happened, but it was one of the first of many crop formations that appear to be drawing our attention to the importance of Venus with highly complex messages about its cycles.

Other significant astronomical cycles that have been shown to be encoded within crop formations include the eighteen-year Saros eclipse cycle and the nineteen-year Metonic cycle of the Moon. Specific dates of lunar and solar eclipses and specific planetary conjunctions have also been found. Researcher Paul Vigay noted that it seems that the geometric language of astronomical time is a particularly favorite subject matter for the creators of crop formations.

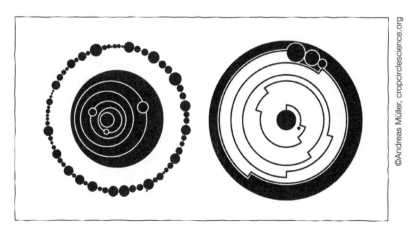

©Andreas Müller, cropcirclescience.org

▲ The missing-Earth and Pi crop formations

Intelligent Life Must Have Created the Pi Crop Formation

Since pi is a universal mathematical constant formed from the ratio of the circumference of any circle to its diameter, once the code has been cracked it can be interpreted by anyone speaking any language anywhere. The pi crop formation has been called the most complex ever found. It is certainly one of the most ingenious. Anyone wishing to communicate in the most universal form possible would do well to match the elegance and simplicity of this geometric solution. When mathematical proofs are demonstrated in cornfields, it seems likely some form of intelligence created them.

Probably the most outstanding crop formation of the 2008 season was found in a barley field at Barbury Castle in Wiltshire. The now famous image of a rotary encoder or ratchet design was at first a seemingly mysterious message, but after more than a week of attempted interpretations, Mike Reed, a retired physicist-engineer from North Carolina, managed to decode it. What he discovered was that the formation represented the value of the mathematical constant pi to a value of ten decimal places. The number is so exact that the tenth digit is even correctly rounded up.

A small dot near the center of the formation represents the decimal point and the numbers are drawn into the crop using 36 degrees of rotation to represent a digit. The first digit, "3," was represented with 3 × 36 = 108 degrees of rotation, while the second digit, "1," was drawn with 1 × 36 = 36 degrees of rotation. The third digit, "4," was drawn with 4 × 36 = 144 degrees of rotation, and so on. "The code is based on ten angular segments, with the radial jumps being the indicator of (the end or start of) each segment," Reed concluded.

80. Messages from Aliens?

Some crop circles have contained messages that specifically claim to have originated from an extraterrestrial intelligence. The first of these appeared in the field next door to a radio telescope at Chilbolton. The formation was instantly recognizable as a clear response to the message to extraterrestrial intelligence that was broadcast from the Arecibo radio telescope in 1974. The message looked strikingly similar, but on closer inspection revealed several significant differences.

The Chilbolton crop formations' reply to the image broadcast by the Arecibo Telescope has been altered so that:

- The human figure has a far larger "alien" head.
- The building block elements of life now include silicon, in addition to the original elements listed.
- There is an extra strand added on the right side of the DNA double helix.
- The diagram that depicts our solar system has been changed to show Earth, Mars, and the moons of Jupiter as being inhabited.

Whoever created this formation certainly went to a great deal of trouble to encode this information. At the bottom, the original

Could Crabwood Be a Hoax?

It's worth noting that the binary code in the Crabwood crop formation is in ASCII, a computer language invented by scientists in the 1960s, which translates into the Western alphabet. No other crop formation has included anything like this. All other binary encoded information relates to dates and astronomical information that is universal and not specific to one language. One of the researchers responsible for decoding this message noted this and the fact that the frame of the picture is uneven and not at exactly ninety degrees, which is unusual considering the precision of the rest of the formation. Thus, he decided that the Crabwood formation was an extremely elaborate hoax.

Arecibo picture had included an outline of the radio telescope itself and its dimensions. In the reply, this had been replaced by a strange multiarmed fractal that would seem to be the aliens' telescope. A much larger version of this image had previously appeared as a crop formation in its own right in the same field!

An even more explicitly alien crop formation appeared at Crabwood in 2002. This contained a giant image of a typical gray alien face, with an elongated skull and enlarged eyes. This was rendered into the crop as a series of lines, much like an old black-and-white television image. The effect was striking and haunting. At the bottom right-hand side of the image was a disc. Read from the inside outwards, the disc contained a binary code that could be translated into the computer language ASCII. The message reads:

"Beware the bearers of false gifts and their broken promises. Much pain but still time. Believe: there is good out there. We oppose deception. Conduit closing."

81. Some Crop Formations Seem Connected to the Mayan Calendar

Crop circles and 2012 seem to be linked in multiple and often mysterious ways. There appears to be a significant link between the numbers expressed in crop circles and those in the Mayan calendar. Researcher Geoff Stray has worked to correlate the geometries of different crop formations with the underlying number harmonics of the Mayan calendar system. These correlations are so extensive that numbers related to nearly all of the known cycles that were being tracked by the Maya have been represented in a type of crop formation. Here are some notable examples of the correspondences:

- **13—Tones:** A thirteenfold star appeared at Huish, Wiltshire in July 2003. This formation is almost identical to a diagram created by Argüellés in *The Mayan Factor* to illustrate the thirteen-baktun cycle.
- **20—Solar Seals:** A Mayan design of a circle with twenty "G" symbols, the Mayan sign for the Milky Way, with thirteen sectors inside the circle, appeared in Waylands Smithy in 2004.
- **33—Symbols in the Tzolkin:** A sun formation with thirty-three divisions in its outer circle appeared in Silbury Hill, Avebury, Wiltshire, in May 1998.
- **52—Calendar round:** A formation with fifty-two circles in the tail appeared in Beckhampton, Wiltshire, in 1998. According to the crop circle researcher known as Red Collie, the Alien face formation at Crabwood in 2002 contains the fraction

of 104/2. This is a key number of the Mayan calendar system expressing the relationship between the calendar round and the Venus round.

- **65—Venus round:** A pentagram with thirteen scales between each pair of arms, making a total of sixty-five scales, appeared in Silbury Hill, Wiltshire, in 2002. The pentagram is the figure traced out by Venus over five of its 584-day cycles, which adds up to exactly eight 365-day Haabs. Sixty-five Venus cycles equal one Venus round, which is exactly the same length of time as two calendar rounds.

- **260—The Tzolkin:** A grid of 780 squares appeared in Etchilhampton, Wiltshire, in August 1997. Three Tzolkins of 260 days each equals 780 days. This is also equivalent to one synodic period of Mars. Crop circle researcher Michael Glickman noted that this formation appeared exactly 780 weeks before the same week in August 2012. Of this he said, "I am convinced that what is being predicted is a dimensional shift . . . which culminates in 2012. By that point, we will fully occupy an entirely different level of being."

- **360—The Tun/Haab:** A formation with twenty outer circles and eighteen inner squiggles appeared in Avebury Trusloe, Wiltshire, in 1999. These numbers represent the eighteen uinals of twenty days each in the yearly Haab calendar.

- **400—The Baktun:** A grid of 400 squares appeared in East Kennett, Wiltshire, in 2000. There are 400 360-day tuns in one baktun.

82. Mayan and Aztec Designs Are Present in Some Crop Formations

Another distinct evolution happened in the complexity of crop formations when Mayan- and Aztec-style designs began to appear around 2001. These incorporate some of the classic motifs of Mayan and Aztec art, like the Mayan "G" shape, which is commonly found on Mayan ceramics and statues. The first of these appeared in Wakerley Woods, Northamptonshire, in 2001 and was named the Aztec Pizza, because it looked like a cross between the famous Aztec sunstone and a pizza. It incorporates eighteen "G" glyphs into its design, which is the Mayan symbol for the Milky Way. Stray has pointed out that eighteen is also a key number in the Aztec sunstone.

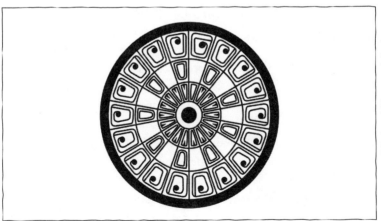

©Andreas Müller, cropcirclescience.org

▲ The Aztec Pizza Mayan Wheel crop formation

Where Crop Formations Appear

Most crop formations are still found in the west of England, and the majority of those are concentrated in the crop circle country of Wiltshire, within a fifteen-mile radius of Avebury stone circle. However, increasingly more complex crop formations are being found on mainland Europe and elsewhere, such as Canada. In 2008, a particularly impressive example was found at Secklendorf in Germany. The formation has a star tetrahedron in the central circle and a complex ring of orbits and planet circles surrounding it. This has been interpreted by Collie as showing near-future orbital locations for Earth, Venus, and Mars on the forthcoming Venus transit of June 6, 2012. It also shows the orbital locations for Earth and Venus on the calendar end date of December 21, 2012.

The evolution of these designs continued to progress over the next few years, reaching a new high point in 2004 with the Mayan Wheel formation that was found at the ancient megalithic site of Silbury Hill, Wiltshire, in August 2004. This formation was considered by many to be the finest of that summer and was widely reported in the world's media, where it was named the doomsday crop circle. This was because it seemed to evoke a resonance with the 2012 end date of the Mayan calendar, although no detailed analysis or interpretation (other than mentioning the formation's Mayan appearance) was offered in the coverage.

The formation does, however, seem to contain a number of symbols that are related to 2012. Firstly, around the edge are double-G glyphs, called jaguar snouts by the Maya, a symbol that represented the entrance to the underworld. This is found at the dark rift at the center of the galaxy, from which the winter solstice sunrise will appear to emerge on December 21.

Some interpreters have suggested that the formation also represents the four completed worlds in the Mayan creation story and the current fifth one. The previous worlds are represented by the outer rings, which are divided into four equal parts by a broad dark space in between. The open circle at its center, surrounded by two wings, represents our current fifth Sun.

83. The 2012 Crop Formation

There are also a number of dates relevant to 2012 that have been decoded from messages in crop formations, including one almost exactly describing the end date of the calendar on winter solstice 2012. It appeared in July 2008, once again near Avebury in Wiltshire, as a more complex version of the missing Earth-type formation. This time it contained the orbits of all nine planets, not just the inner ones. The planetary alignment that the formation points to is almost exactly the Mayan calendar end date of December 21, 2012. There is some debate about the precise date that is being described; some researchers think it is a slightly better fit for the positions of the planets on December 23. Nonetheless, this is still remarkably close and the clearest connection between 2012 and the crop circle phenomenon that has so far been recorded.

All nine planets appear there precisely as they will be located in space at the winter solstice, with one exception—Pluto, which is now considered to be a dwarf planet. While the positions of the inner terrestrial planets are marked with a small doughnut and the outer

2012 and Plasmic Energy

The theoretical link between plasma vortexes and the creation of crop formations is thought to be attributable to an increase in the amount of available plasmic energy in our solar system. We may be seeing the culmination of a process of cosmic evolution much longer in the making than the very rapid one mirrored in the summer fields. Study of UFOs and crop circles has been largely relegated to the paranormal fringes and extremes of science, but now the common thread of plasma may link them to a range of vitally important subjects from climate change on Earth to much greater cosmic weather patterns throughout our galactic neighborhood. The year 2012 may turn out to be a catalytic point, when all these phenomena come together in a more integrated understanding of our place in the universe.

gas-giant planets with a large doughnut, Pluto is marked with a medium-sized ring, which immediately brings attention to it. It is also thirty degrees—or about twenty years of its orbit—ahead of its expected location. This has led to lots of speculation about what this could mean. A favorite explanation is that the orbit of Pluto has somehow been disrupted by the gravitational influence of a body like the legendary Niburu, sometimes called Planet X, returning to our solar system. The name Niburu comes from the ancient Sumerian and describes a planet that is hypothesized to exist on an extremely long orbit, only entering our solar system every few thousand years. The arrival of Niburu, the wanderer, is seen as a time of great change. The return of such a planet would likely cause significant disruption to any other planet whose orbit it came close to.

84. Could Planet X Arrive in 2012?

One persistent source of speculation about 2012 is that it heralds the return of a wandering and possibly destructive planet or star to our solar system. This is sometimes referred to as Planet X or Niburu.

The quest to find the hypothetical Planet X has a long and colorful history in astronomy. The American astronomer Percival Lawrence Lowell first popularized the idea that there was a planet beyond the orbit of Neptune at the start of the twentieth century. He believed that discrepancies in the orbits of Uranus and Neptune could be accounted for by the existence of a large, yet-to-be discovered planet located beyond the orbit of Neptune.

This followed in the footsteps of a long tradition of planetary discovery by studying the irregularities in the orbits of the outermost planets. It was perturbations in the orbit of Saturn that led William Herschel to search for and find the planet Uranus. In 1843, the mathematician John Couch Adams discovered perturbations in the orbit of Uranus. In 1846, this led to the discovery of Neptune by Johann Gottfried Galle of the Berlin Observatory.

In 1930, it seemed that Lowell had been vindicated by Clive Tombault's discovery of Pluto. Tombault was hired by the Lowell Observatory to examine thousands of photographs of the night sky, continuing the search for Planet X. He found the new planet just six degrees from one of the two positions that Lowell had suggested as likely locations for the discovery. The new planet was named Pluto partly because it contained the initials of Percival Lowell. The

Other Objects Near Pluto

Pluto's status in the solar system has declined in recent years. Much smaller than any of the other planets and not much bigger than the largest of the asteroids, astronomers began to question whether Pluto should be considered a planet at all. As the power of telescopes rapidly increased during the 1990s, more trans-Neptunian objects (TNOs) of a similar size as Pluto were discovered just beyond its orbit in what is known as the Kuiper Belt. Named after astrophysicist Gerard Kuiper (rhymes with "viper"), the Kuiper Belt is an area of space on the edge of the solar system near the ecliptic plane. The Kuiper Belt extends some 5 billion miles from the sun, a little more than fifty times the distance between Earth and the sun. More than 800 Kuiper Belt objects have been found.

astronomical symbol for the planet is a combination of the letters 'P" and "L" in recognition of this fact.

In 1978, the mass of Pluto was found to be only 60 percent of that of Earth's moon, far too small to be the perturber of Neptune's orbit, showing that it couldn't be Planet X. In 1993, data analyzed by the astronomer Myles Standish from Voyager 2's fly-by of Neptune in 1989 showed that Neptune had a mass that was 0.5 percent smaller than had previously been calculated. When that change in mass was taken into account, the supposed discrepancies in the orbit of Uranus disappeared. This effectively ended the search for another large planet near the orbit of Neptune.

85. Nemesis Could Be Responsible for Previous Extinction Events

Another idea that has been gaining a lot of momentum is the theory that the sun has a yet-to-be discovered companion star. Richard A. Muller dubbed this hypothetical companion star Nemesis. Nemesis is theorized to be orbiting the sun at a distance of between one and two light years.

The existence of this star was originally posed to explain an apparent cycle of mass extinctions in the geological record. In 1984, paleontologists David Raup and Jack Sepkoski published a paper claiming they had identified a statistical pattern in species extinction rates over long periods of time. Using evidence from the fossil record, they were able to identify twelve major extinction events over the last 250 million years. The average time interval between extinction events was 26 million years, and two of these events were shown to coincide with large impact events. Raup and Sepkoski were not able to identify the cause of this, but they suggested there might be an astronomical explanation.

In response, Muller came up with the idea of Nemesis, named after the Greek goddess of retribution, as a death star whose orbital interactions with our solar system periodically caused showers of comets to be thrown toward the inner planets. The more Muller examined the hypothesis, the more plausible it seemed: During the passage of Nemesis through or near the Oort cloud, the star's gravity would dislodge millions or even billions of comets from their once-stable orbits. These comets would then head toward the inner solar system, pulled in by the sun's gravity. The few that inevitably collided with Earth would be likely to result in mass extinction events.

The exact nature of Nemesis, if it exists, has yet to be determined. Muller's preferred candidate is a common red dwarf star that would

What Are Binary Systems?

More than half of all stars in the galaxy are part of a binary system. This means they are created when two or more stars are born out of the same interstellar dust cloud. Binary stars orbit around a shared common center point in space. The smaller of the two stars has the larger orbit. So, you may be wondering: If the sun has a companion star and we are part of a binary system, why can't we see the other star? The answer: If it is a red dwarf, it has probably already been discovered and catalogued but simply not recognized as a companion to the sun. Its proper motion, or movement across the sky, may be very small, as it may be moving away from or toward us. If it is a brown dwarf, it may be relatively close but still difficult to detect visually. The next generation of orbiting telescopes may be able to discover it.

be visible through binoculars or a small telescope. More than 3,000 red dwarfs that fit into this category have been cataloged, but their exact distances are not yet known. Red dwarfs are the most common type of star found in the galaxy. They are small and relatively cool in comparison to our sun.

Other astronomers argue that the unseen companion to the sun is much more likely to be a brown dwarf. These are unborn stars that briefly ignite but then quickly die. The initial ignition dust cloud that surrounds them makes them very hard to see and even the most powerful telescopes could miss such a star, even if it were very close to our solar system. Brown dwarfs are very common, and if one were orbiting somewhere around the outer edge of the Oort cloud it would be a good fit for the perturber sending comets toward us.

The astronomer P. R. Weissman considers that the likely mass of the perturber means it could be a black dwarf. These are hypothetical stellar remnants, created at the end of the life cycle of a normal star when the processes of nuclear fusion shut down. It then no longer emits significant heat or light. The time required for a star like our sun to reach this state is calculated to be more than 13.7 billion years, longer than the current age of the universe.

86. The Secrets of the Oort Cloud

The Oort cloud is a distant conglomeration of frozen objects that surrounds the solar system between 900 billion and 4.5 trillion miles from the sun. The edge of the cloud is thought to be a massive 1.5 light years from the sun, more than a third of the way to the nearest star to the sun, Proxima Centauri. Most comets are thought to originate from this region. These comets spend millions of years in the Oort cloud until they are deflected into an orbit that takes them into the inner solar system where we can see them.

Measurements of the regularity and distribution of comets coming from the Oort cloud have shown patterns that cannot be accounted for by the gravitational influence of our galaxy, known as the galactic tide. It is the edge of the Oort cloud, trillions of miles away from the inner solar system, that is currently the subject of much study and many speculations, some of which are very relevant to life on our planet.

Dr. John Murray of the U.K.'s Open University suggests that there is another planet in deep space at the edge of the Oort cloud. After studying the motions of long-period comets, Murray has detected what he believes are telltale signs of a single massive object that has deflected all of them into their current orbits. He calculates that there is only a one in 1,700 chance that this is a coincidence. In a research paper published in the *Monthly Notices of the Royal Astronomical Society*, he has suggested that the so-far-unseen planet is a supergiant, several times bigger than the largest planet in the solar system, Jupiter.

The orbit of this supergiant would be approximately 3,000 billion miles (32,000 times farther away than Earth) and it would take as many as 6 million years to orbit the sun. Such a distant planet would be slow-moving and difficult to detect despite its huge size. What is probably the most interesting part of Murray's hypothesis is the remarkable suggestion that the planet orbits our sun in the "wrong" direction, counter to the orbits of all the other known planets. This has led Murray to believe that it did not form in this region of space and that it could be a planet that "escaped" from another star. He has even been able to calculate that the predicted supergiant lies in the constellation of Delphinus.

Long-period comet
Comets whose orbits are longer than 200 years.

87. Is the Sun Orbiting Another Star?

The yet-undiscovered binary companion to the sun may also be responsible for producing the effect of the precession of the equinoxes. This would link the star to the galactic alignment of the 2012 era, when the winter solstice sun can be seen to rise in conjunction with the center of the galaxy. Redefining precession as a heliocentric phenomenon based on the motion of the sun around another star would also revolutionize the science of astronomy in a shift equal to that caused by the theories of Copernicus.

Heliocentric
Having the sun as the center.

Copernicus's explanation of the cycle of precession was part of a revolutionary new model of our solar system that moved the center of the known universe from Earth to the sun. Ironically, the motions described by Copernicus to create the modern heliocentric or sun-centered model of the solar system are all actually geocentric or Earth centered:

- The first motion of Earth proposed by Copernicus was the rotation of the planet on its axis to produce the twenty-four-hour day.
- The second motion was the orbit of Earth around the Sun that created the 365-day year.
- The third motion was the wobble of Earth on its axis to produce the effect of the precession of the equinoxes. A century after Copernicus introduced his theory, Sir Isaac Newton suggested that this was an effect caused by the combined gravity of the Sun and the Moon.

In Copernicus's model, the sun doesn't move. However, we now know that the sun orbits the galactic center and oscillates

2012 PERSONALITY:
Walter Cruttenden

Walter Cruttenden describes himself as an amateur archaeoastronomer. After a successful career in finance, he has dedicated himself to promoting and refining the binary star theory of precession. In 2001, he founded the Binary Research Institute (BRI). He has written a book on the subject called *Lost Star of Myth and Time* and made a documentary film called *The Great Year*. The BRI also holds an annual Conference on Precession and Ancient Knowledge (CPAK).

up and down in relation to the plane of the galactic equator. The binary star theory introduces heliocentric motion, the idea that the solar system is moving in a curve through space in relation to its companion. This, it suggests, is the cause of precession, not Earth wobbling.

One of the great successes of the Copernican heliocentric solar system was to remove the need for epicycles. In the Earth-centered worldview founded by the ancient Greek astronomer Ptolemy, this complex idea was used to describe the strange looping motions seen in the planets as they traversed the sky. Similarly, the binary star model dispenses with the need for the complex and fragmented mathematics required to support the lunisolar theory of precession.

The binary star theory asks us to take an imaginative leap that is equal to the one that Copernicus asked the medieval world to take. At the time, Copernicus's theory had not only scientific but also theological consequences, as the world was to lose forever its unique place at the center of creation. In a similar way, the binary star theory of precession removes the sun from the center of the modern worldview.

Lunisolar

Relating to the relationship between the moon and the sun.

PART FIVE

What Happens in 2012?

This book has explored many of the numerous theories concerning what may happen to the world as we know it in 2012. Many philosophers and scientists have attempted to bring these very different ideas together with the goal of explaining and spreading awareness of the possible changes to come.

In this final part, you'll find all the insights and information integrated into some suggestions about how to prepare for the highly anticipated end point of the Mayan calendar.

88. 2012 May Be a Global Tipping Point

The changes in both global society and our physical planet can be studied just like any other large, complex system. Over the last twenty to thirty years, two branches of science have increased our ability to understand behavior in systems with a large number of variables: complex systems theory and chaos theory. The laws that these theories articulate work just as well for any large system, from the swarming of locusts to global social trends.

Ervin Laszlo is a former professor of philosophy, systems science, and future studies and has been nominated three times for the Nobel Peace Prize. He is also the founder of the Club of Budapest, an international think tank dedicated to positive global change, whose members include the Dalai Lama and Mikhail Gorbachev. In his recent book *The Chaos Point*, he has used ideas from both chaos theory and complex systems theory to propose that in late 2012 we will reach a global tipping point.

Chaos theory demonstrates that complex systems tend to alternate between periods of dynamic stability and abrupt phases of critical instability. When this critical instability becomes irreversible, this is a sign that a bifurcation point is about to be reached. As this happens, the old system appears to collapse and the values and patterns that have dominated the way things operate begin to break down. At first, this looks like a catastrophic failure. From a broader perspective, however, the breakdown can provide an opportunity for a breakthrough. Historical examples could include the fall of the Berlin Wall or the stock market crash of 1929. The collapse of one thing is often just the beginning of another.

When the old order falters, there is a momentary pause. One set of values no longer applies, but nothing has yet taken its place. This usually brief state can seem to be either chaos or freedom, depending on the way it is viewed. Complex systems theory has shown that when systems reach a sufficient level of organization and complexity, they tend to reorganize themselves in a higher level of order after a collapse. This pattern applies whatever the parts of the system are, from atoms to people.

There are many signs that this sort of process is currently at work in global society. The worldwide banking and financial crisis of 2008 is an example of the sort of oscillation that typically precedes a bifurcation point. Share and commodity prices rise and fall in a cycle of increasing volatility. Large fluctuations may be followed by periods of relative calm or consolidation, but the underlying trend is irreversibly

A Gateway to Something Better or Worse?

Laszlo predicts that a series of underlying social and economic trends will force society to make a choice between a global breakdown and a global breakthrough. In *The Chaos Point*, he concludes, "2012 is indeed likely to be a gateway to a different world, but whether to a better one or to a disastrous one is yet to be decided." The result depends on the new set of initial conditions. These are fundamental changes in structure that will end the increasingly volatile fluctuations inherent in the old order.

leading toward a chaos point. At the chaos point, the butterfly effect applies: Small changes in the new initial conditions of a system can create enormously different outcomes.

Laszlo suggests that the choices that we all make around the chaos point of 2012 will have lasting consequences for generations to come.

89. The Noosphere and Humans' Collective Consciousness

Ukrainian geochemist Vladimir Vernadsky first coined the term "biosphere" to articulate the concept that all of the living systems on the planet could be thought of as a single entity. The biosphere includes the crust of Earth, the landmasses and oceans, and the parts of the atmosphere that life inhabits.

This idea allows the totality of life to be described as a series of processes transforming energy into different forms. Vernadsky's biosphere could then be studied in the same way as an industrial or chemical process, with inputs, outputs, and byproducts. For Vernadsky, the biosphere was "the single greatest geologic force on Earth, moving, processing, and recycling several billion tons of mass a year. It is the central subsystem of a centralized cybernetic system, Earth, which tends towards a dynamic disequilibrium and tremendous internal diversity."

Vernadsky also proposed the idea of the noosphere. Derived from the Greek word for mind, *noos*, it means "sphere of thought." Vernadsky theorized that the noosphere was the third in a series of fundamental transformations that have happened to Earth:

1. The first was the creation of the geosphere, the inanimate, mineral Earth.
2. The geosphere was then completely transformed by the arrival of life and the creation of the biosphere.
3. The biosphere is in turn being transformed by human thought, or the creation of the noosphere.

Vernadsky's theory was that the noosphere truly emerges when humanity masters the science of the physical realm and is able to reorganize the primary sphere of matter. Marked by the beginning of the atomic age, the noosphere is still considered an extremely significant theory in the mainstream of post-Soviet science.

Vernadsky's original idea of the noosphere was intended to be a material,

2012 PERSONALITY: Teilhard de Chardin

Chardin is, in many ways, the archetypal philosopher of 2012. He formulated the fundamental dynamic behind his philosophy into a principle called the law of complexity/consciousness. The law states that matter shows an inherent compulsion toward complexity. Chardin proposed this as a universal model of evolution, where everything is becoming more complex. The emergence of humanity takes this process to a higher plane. Humans' self-awareness creates a new layer of consciousness: the noosphere. With the advent of the noosphere and self-consciousness, the processes of evolution become increasingly voluntary.

scientific description of the process of human civilization transforming the environment that gave it life. This idea was then taken
up by the Jesuit philosopher Teilhard de Chardin, who developed it
into something much more expansive. Chardin saw the noosphere
as the perfect way to describe the emerging collective consciousness
of humanity.

Unshackled by the constraints and rigors of science, Chardin
was looking for a way to describe what he saw as the process of a
self-reflective consciousness emerging in humanity and spreading
itself across the planet.

90. The Omega Point Is the Foundation for Many 2012 Theories

The culmination of the process of evolution ends in what Chardin
called the omega point, the point at which the complexity of the universe has reached its maximum function and has become organized in
the most optimal way possible. The omega point is nothing less than
the purpose of history, the agonies and ecstasies of which are redeemed
in a single moment of supreme meaning. Other 2012 theories, like
McKenna's vision of a transcendent moment of infinite novelty at the
end of time, owe a great deal to Chardin's idea of the omega point.

The processes of history are normally figured to start with a
defined moment, like the big bang of modern physics or the biblical
seven days of creation. Cosmologies organized in this way are called
ontologies. In Chardin's cosmology, it is the other way around. The

omega point is the strange attractor drawing us toward it. Belief systems that are based on a defined end point rather than a start point are called teleologies. The five attributes of the omega point are:

1. Already existing. This attribute explains the power of the omega point to draw us to itself.
2. Personal. The omega point must complement and integrate individuality rather than annihilate it. Otherwise, the law of complexity/consciousness is broken.
3. Transcendent. It must not be a product of the universe, but a preexisting condition from which the universe arises.
4. Autonomous. The omega point is not subject to the laws of space and time.
5. Irreversible. It must be attainable and permanent.

At the omega point, Chardin believes that the perfected noosphere becomes synonymous with the "Christosphere," or collective Christ consciousness of humanity. Chardin's philosophy of the noosphere, despite his rejection by the Catholic Church in his lifetime, is a profoundly religious, if unorthodox, vision. In the perfection of the noosphere, humanity redeems itself and achieves transcendence.

91. The Telepathic Human: Coming in 2012?

Chardin's omega point specifically envisages thought-to-thought telepathy between humans becoming the norm as our communication networks evolve toward the omega point. Other 2012 theorists have picked this up. Argüellés has recently written about the emergence of a new species of telepathic human, *Homo noosphericus.*

Argüellés has spoken about the emergence of *Homo noosphericus* as being a permanent physical change that is imminent for humanity. "We are not talking about being the same as we are now," he says in *Cosmic History Chronicles, Volume 3.* "We are talking about a full-on mutational shift, a frequency shift in which our atoms and molecules begin to vibrate at a higher frequency and, thus, genetically self-correct." After this shift, according to Argüellés, the linear third-dimensional perceptions of history will be replaced by a total holographic comprehension of every moment of every day. We will then be able to truly view ourselves as multidimensional beings. From this new perspective, we will be able to create and perfect alignment with our fourth- and fifth-dimensional galactic-celestial bodies.

Argüellés has talked about the technosphere as a necessary intermediate stage in the biosphere-noosphere transition described by Chardin. The technosphere could be defined as the sum of all the connecting parts of the global technological society that cover the biosphere like an electronic sheath. In this model, outlined in his book *Time and the Technosphere,* technological society, especially the Internet, is temporary scaffolding that could be used to trigger the genesis of *Homo noosphericus.* Ultimately, Argüellés sees material technology as an alluring trap that must eventually be discarded in order to restore our original connection with nature. The coming telepathy of *Homo noosphericus* represents the real Internet. This is essentially just a development of Chardin's ideas and is in total agreement with his teleological philosophy.

92. 2012 Could Signal the Arrival of the Global Brain

Chardin's ideas are further developed in the work of the author Peter Russell. In his 1983 book *The Global Brain*, Russell puts forward the idea that the true purpose of humanity is to evolve into a massively networked global brain for our planet. In Russell's theory, each individual would perform a function similar to an individual cell in the brain.

By learning to think in this networked way, the global brain would represent a mind of its own with unfathomably large computational power. Humanity would then also be able to act in a coordinated and harmonious manner and avoid the conflict, greed, and exploitation that has characterized

2012 . . . The Start of the Psychozoic Era?

Vernadky believed the next geological era of Earth would be called the Psychozoic Era or the era of mind. The theories of Chardin and Vernadsky both seem to provide valuable intellectual tools to help grasp the end-times zeitgeist of 2012. As the writer Peter Russell wrote in the book *The Mystery of 2012*: "2012 is a symbol of the times we are passing through. It represents the temporal epicenter of a cultural earthquake, whose reverberations are getting stronger day by day." If that is the case, the current time may one day be thought of as the Lower Noospheric Era, to reflect the significance of the discovery that humanity is part of one living Earth. The era when humanity collectively acts on that discovery may then be known as the Upper Noospheric Era. It may well be that 2012 indeed represents the epicenter or threshold between the two.

its history. Russell's timeline for the emergence of the global brain coincides with 2012, which he views as a "white hole" in time, his version of Chardin's omega point.

In support of this idea, Russell cites the fact that it takes approximately ten to the power of ten (or ten billion) atoms to form the most basic level of unicellular bacterial life. This number seems to him to be a necessary minimum for the sufficient complexity for the evolution of life. In a parallel to the evolution of life from matter, Russell suggests that a similar number of brain cells (in the region of ten billion) in the neocortex are required to produce the reflective consciousness characteristic of humanity. If this turns out to be a general principle of evolution, Russell suggests that the next evolution could be represented by something in the region of ten billion minds beginning to work together as one global brain.

Russell formulated these ideas in the early 1980s, and correctly predicted the importance of emerging computer networks in an information revolution. Russell noted that by 1900, more people were employed by industry than the previously dominant activity of agriculture. By the mid-1970s, the number of people engaged in the processing of information (in all of its aspects, from publishing to banking to media to all computer-related occupations) had caught up with those engaged in industry—the processing of energy and matter.

"From that time on," Russell declares, "information processing has been our dominant activity." We had entered the Information Age. The Industrial Revolution took 300 years. The Information Age has been in ascendancy for thirty. Russell predicts that this fits into a pattern of ever-accelerating evolution that stretches back to the dawn of life on Earth:

- The first simple life forms evolved 4 billion years ago.
- Multicellular life appeared about 1 billion years ago.
- Vertebrates with central nervous systems developed several hundred million years ago.
- Mammals appeared tens of millions of years ago.
- The first hominids appeared a couple of million years ago.
- *Homo sapiens* appeared a few hundred thousand years ago.
- Language and tool use developed tens of thousands of years ago.
- Civilization, the movement into towns and cities, occurred a few thousand years ago.
- The Industrial Revolution began three centuries ago.
- The Information Revolution is a few decades old.

Russell's Wisdom Age

Our current computer technology has created the ability to communicate and transfer more information than ever before. In Russell's model, this creates the opportunity for cultures from around the world to share insights about enlightenment and different wisdom traditions. Teachings that otherwise would only have been available to a tiny minority are now effectively available to everyone. Thus, Russell predicts the dawning of a Wisdom Age. According to Russell, the exponential growth of the Wisdom Age will be so rapid that it will outstrip the growth of the Information Age in the very near future. "Because each new phase of evolving intelligence takes place in a fraction of the time of the previous phase, we can expect the dawning of a Wisdom Age to take place in years rather than decades. It will be standing on the shoulders of the Information Age."

Similar to the theories of Calleman and McKenna, Russell believes that cultural acceleration is leading toward a culmination at some time in the very near future. Although he is not attached to any specific prediction for what will or won't happen on December 21, 2012, he does embrace the general principle of a major evolutionary breakthrough for humanity somewhere around that time.

93. The Theory That Computers Will Become More Powerful Than Humans

Not everyone who is predicting an impending singularity in human development is suggesting that it will be a spiritual Renaissance. Futurists and technologists are predicting that we may be heading toward a technological singularity. This idea was first proposed by the mathematician Vernor Vinge. Vinge based his idea on Moore's Law, the observation that the rate of increase in computing power has been consistently exponential for the last fifty years.

Computing power now doubles approximately every eighteen months. If this trend were to continue, Vinge and others argue, computers within a decade of our current era will be more powerful than the human brain. Once this happens, they suggest it is likely that computers themselves will take over the designing of future computers. This could then lead to a runaway train kind of scenario where machines rapidly become much smarter than the humans who initially created them.

In this version of convergence, the technosphere would be the goal, not just the means. The technological singularity would be a convergence of all technologies, until humans became totally embedded and submerged into a virtual world. This singularity is a state in which humans will be components of a cybernetic social network of such complexity that no one person will be able to understand more than a tiny fraction of the whole.

One of the chief proponents of this idea is the futurist Ray Kurzweil, who describes this near-imminent technological omega point in his book *The Singularity Is Near*. His vision of the future is not just a matter of his opinion; it is a template that is being acted upon and implemented by the multinational corporate world.

Kurzweil predicts that not only will computers become more intelligent than humans, but also that computers will become so powerful that it will be

2012 PERSONALITY: Ray Kurzweil

Kurzweil believes that his only responsibility is to stay alive until this technology is in place so that he can live forever. Consequently, he is taking a large number of health supplements and is planning his life to avoid taking any unnecessary risks. Kurzweil's current estimate is that computers will surpass the power of the human brain sometime in the next decade and that the singularity will occur sometime around 2045.

possible to download the entire contents of the human brain into one. The result of this would be a sort of digital immortality. Kurzweil does not see a problem with consciousness being transferred along with the information from a brain.

94. Or, Could Humans' Capabilities Grow?

Transhumanism is a movement that supports the use of science and technology to radically alter and improve human mental and physical characteristics and capacities. Transhumanists regard mortality as just another technical challenge that will eventually be overcome when scientific advances make immortality possible.

The term "transhumanism" was first coined in 1966 by FM-2030 (formerly known as F. M. Esfandiary), a futurist who taught new concepts of the human at The New School in New York City. The term was aimed at describing those who believe technology and science will eventually have a solution for all of humanity's problems. For most transhumanists, the technological singularity is an important goal.

Transhumanism is sometimes associated with the term "H+," which stands for "human enhancement." H+ symbolizes the belief that by using technology, humanity can grow capabilities far beyond what is currently possible. The abilities of H+-designated humans may be extended so far in terms of intelligence, longevity, and even strength that they will be considered posthuman.

Some of the most extreme examples of the idea of technological convergence can be found in the ideas of Dr. Frank Tipler. Extrapolating from the known laws of physics, he points out that if the universe reaches a finite point of expansion and then begins to contract, it is likely to end in a big crunch. This is an almost exact reverse of the big bang. At this point, all of the matter in the universe will have converged on one point and there will be zero available free energy.

Tipler thinks this situation can be redeemed by the possibility of creating a truly giant computer that will model the entire universe. Even though the length of the universe is finite (and the computer is part of that universe), the computer's ability to process increasing quantities of information is theoretically infinite. According to Tipler, this will result in a transcendent omega point, as the computational capacity of the universe will be accelerating exponentially as time runs out.

The Edge of Eschatology

Kurzweil's vision of the world post-singularity is strikingly similar to Tipler's. Kurzweil predicts that human-created artificial intelligence will eventually wake up all the matter of the entire universe into a state of consciousness. These sorts of concepts, along with those of Terence McKenna, form the far edges of eschatology, the study of the end times. The year 2012 is the archetypal convergence point for all end-times theories. These various ideas are currently now competing for what the philosopher A. N. Whitehead called the formality of actually occurring.

95. How Likely Is a Global Crisis Event in 2012?

Some of the possible events that have been covered in this book are relatively unlikely to occur in 2012, or even shortly after. The search for discovering the real message of 2012 has meant considering apocalyptic outcomes (such as a physical pole shift or a major galactic superwave event) as possibilities, but even the founder of the galactic superwave theory, LaViolette, is relatively optimistic. If there is an event, he believes it is likelier to be a more minor wake-up call type of event, rather than a cataclysm.

That one of these types of major event will happen at some time in our planet's future is actually very likely, but the time scale may be vast. Superwave events of some magnitude may happen once every 13,000 years. Magnetic pole reversals happen, at most, once in 125,000 years. A physical pole shift has probably happened only once in our planet's history, so even a generous guess would put that probability in the order of once every few hundred million years or so.

One of the more likely major events with far-reaching global impact would be the possibility that the Oort cloud perturber, whatever that turns out to be, sends more comets into the inner solar system. However, a cometary impact happening in exactly 2012 is very unlikely.

According to the theory pioneered by paleontologists Raup and Sepkoski, a large event happens approximately once every 26 million years. Even if that were timed to coincide with the galactic alignment marking the end of a 26,000-year precessional cycle, the odds would still be 1,000 to one. Raup and Sepkoski calculate that we are not due for another extinction-level event for perhaps another 20 million years.

What is more likely than any single apocalyptic event in 2012 is that 2012 will mark a major watershed moment in several very important incremental changes. In the year 2012, the fact that climate change, oil consumption, and the human population are all increasingly accelerating may become much more obvious. Major changes to our lifestyles may be required because of the impact of these incremental changes.

96. What Will Replace Oil as a Global Energy Source?

Peak oil is the term that describes the point at which the global production of oil is maximized and we have effectively depleted half of all the recoverable oil reserves on the planet. Most estimates put peak oil as happening before 2012. This is a major landmark in the history of

2012 PERSONALITY: Patrick Geryl

Learning about some of the more extreme possibilities of 2012 makes it tempting to adopt a survivalist mentality and head for the hills. In *How to Survive 2012*, Patrick Geryl writes that in 2012 a pole shift will result in a mile-high tidal wave. He suggests the only survivable options are to either be more than 4,000 feet high in a mountain range in an underground bunker or to be in a special unsinkable ship. Unfortunately, this kind of preparation only suffices for this very unlikely type of global catastrophe, and even then it is no guarantee of survival. In a broad range of much more likely scenarios, isolated survivalists are actually less likely to survive.

global technological civilization, which is still massively dependent on oil as its major resource base. Shortly after this point, reserves of natural gas will also hit their production peak and go into decline. It is impossible to talk about what may happen in 2012 without assessing this as a factor, as it may be a dominant theme.

Possible consequences of peak oil include:

- As supply dwindles, energy prices inevitably increase substantially.
- Industries that are based on the cheap availability of energy or rely heavily on importing or exporting will be the hardest hit.
- Shortages of consumer products may follow as the economy is forced to retool.
- Economic crisis may be exacerbated by speculation on wildly varying energy prices, making long-term planning extremely difficult for even vital and basic industries.
- Oil-dependent food production will be badly affected and food prices are likely to increase substantially.
- Energy-intensive health care is likely to be very seriously impacted by a global downturn based on escalating energy costs.
- Treatments for diseases and illnesses that require significant resources could be made economically nonviable. Health insurance schemes and health services will risk collapse and bankruptcy.

The fundamental problem that global industrial civilization now faces is that it has grown accustomed to the richest source of stored energy ever found. Many people believe that oil will be seamlessly replaced by new or alternative technologies, like new nuclear or wind and wave power. Unfortunately, none of these power sources, or even all of them combined, is capable of replacing oil. This is because oil's net energy—the energy cost of harvesting it against the power it generates in its lifetime—is so enormously high. The net energy stored in sweet light crude oil is around 200:1.

By contrast, the best alternative technology is the modern windmill, which is around 6:1. Some estimates of photovoltaic cells estimate their net energy to be just slightly more than 1:1. This means that new technology, unless it improves its net energy value by at least tenfold, will not replace the easy, cheap energy we are accustomed to now.

97. How to Prepare for "The Long Descent"

The end of the current form of civilization is not necessarily going to be a rapid affair, occurring on one day or even one year. In *The Long Descent*, author John Michael Greer estimates that on average, civilizations take around 250 years from the onset of collapse to complete dissolution. Greer observes, "The process of catabolic collapse unfolds, in a stair-step process alternating periods of crisis with breathing spaces at progressively lower levels of economic and political integration."

Rather than a sudden apocalypse, Greer believes we should anticipate a long descent toward a different kind of society based

Catabolic collapse

John Michael Greer's definition of what happens to a civilization when it can no longer meet the demands for resources and energy to support itself. It then begins the process of feeding on its own parts.

on vastly lower energy consumption. He has called this the deindustrial revolution. The likelihood is that a move toward lower energy consumption, whether forced or voluntary, will result in significant political upheaval. In Greer's opinion, "Many of today's political institutions will not survive the end of cheap energy and the changeover to new political arrangements will likely involve violence." Less developed countries that still have substantial subsistence agriculture and less dependence on fossil fuel energy will be better placed to cope. Here are some suggestions from Greer's book for coping with the changes:

Facet	Solution
Declining energy availability	Reduce energy use
	Plan on reducing energy usage by at least half
	Practice dealing with power blackouts and be prepared to do without power at all, if necessary
Economic contraction	Choose a viable profession
	Choose to learn a craft that requires modest energy inputs and is going to be of use in the local production of essential goods and services

Facet	Solution
Collapsing public health system	Take charge of your own health
	Learn about preventative medicine and other alternative forms of treatment
	Take an advanced first aid class and have necessary medical supplies on hand
Political turmoil	Practice community networking
	Get involved in local community groups, such as local farmers markets, community gardening groups, and social forums

98. Try Practicing Coherence

Coherence is a measurement of how the body and brain are functioning together as one integrated unit. Coherence registers in the rhythms and waveforms of the heart, which in turn has the effect of bringing the brain, respiratory, and other systems into synchronous alignment. In a state of coherence, very little energy is wasted and the combined systems of the body and mind work together in maximum efficiency.

The heart generates an electrical field that is many times stronger than the one the brain produces. This field extends out from the body and can be measured up to several feet away. Researchers at the Institute of HeartMath, an international nonprofit education and research organization, have shown that, in a state of heart coherence, a person's electrical field can have the effect of almost synchronizing the brain waves of people they come into close proximity with to a more coherent state. This shows a scientifically measurable connection that is not dependent on touch.

When a person is in coherent alignment, it is possible to pass that coherence on to others by sharing activity such as a like-minded project or team sporting activity. If a group of people go into coherence together, it is self-reinforcing, resulting in higher and more stable states of coherence.

The Institute of HeartMath believes it is possible to extend states of coherence out into the world to create profound healing effects and harmonious social change. By consciously developing and maintaining states of personal and community coherence, we are then able to have a positive impact on everyone we meet.

Developing more coherence is easier than it sounds. The Institute of HeartMath gives a simple outline for an exercise that can be done daily for around five minutes. Personal benefits are claimed to include an increase in focus and effectiveness. According to Heart-Math, the exercise is also beneficial for the planet, even if the person practicing is not in total coherence. This exercise can be practiced while doing ordinary daily activities and does not require a special time to be set aside. Information is available at *www.heartmath.org*.

Here are some simple daily exercises to build coherence:

1. Breathe and calm yourself in whatever ways you choose.
2. Choose something or someone you appreciate and radiate the feeling of appreciation to them for about two minutes. (This helps open the heart more and increases your effec-

tiveness when you start sending care to the planet or to a situation in need.)

3. Now evoke genuine feelings of compassion and care for the planet.
4. Breathe the feelings of compassion and care going out from your heart.
5. Radiate the genuine feelings of compassion and care to the planet or to a specific area of immediate need.
6. See yourself, along with other caretakers, participating in a process of healing and facilitating peace.

99. Everyday Ways to Prepare for 2012

You can prepare for many of the forecasted changes in ways that do not necessarily disrupt your normal everyday life. Even gradual moves toward self-sufficiency and sustainability are helpful. These can range from reducing your

2012 PERSONALITY: Maharishi Mahesh Yogi

Maharishi Mahesh Yogi, the founder of Transcendental Meditation, predicted in 1960 that if a critical mass of around 1 percent of a community were to practice this form of meditation, a measurable improvement to the quality of life for the whole community would be recorded. In 1974, a study showed that, on average, if 1 percent of a community was practicing the transcendental meditation program, the crime rate would fall by around 16 percent. This has now become known as the Maharishi effect. These studies have been repeated over the following decades on several different continents and with large groups of meditators. The average drop in crime across all these studies was around 11 percent.

consumption of energy and nonrecyclable resources to planting vegetables in a community garden. Here are options to consider:

- Participate in a transition town, a network of communities that is planning for the postcarbon economy, including the creation of alternative local currencies.
- Explore permaculture, the creation of integrated urban and rural design solutions for high-yield, low-maintenance organic agriculture.
- Adopt a raw-food lifestyle: adopting a more organic and closer-to-nature diet can be helpful in staying grounded and healthy through accelerating change.
- Join the Global Ecovillage Network, a worldwide alliance of intentional communities for those who are more committed to a fundamental change in lifestyle.

Meditation is another practice that is thought to be very important in preparing for 2012. In *The Global Brain*, Peter Russell has developed a five-part meditation program that is available on double CD. It is designed for coping with the challenges of embracing the shift in values that will occur with the emerging 2012 paradigm.

2012 Mindshift Meditations
- **Presence:** Finding peace in the moment
- **Befriending discomfort:** Working with difficult feelings and rigid attitudes

- **Inner wisdom:** Tapping the guidance that waits within you
- **Loving kindness:** Developing greater compassion and community
- **Clarifying purpose:** Strengthen your life's vision

These meditations are intended to help a person stay grounded and remain composed, no matter what the changes bring.

100. Documentaries and Websites about 2012

Many online forums discuss how best to embrace the changes of 2012. Two very useful points of focus for this community are the multichannel blogging website Reality Sandwich (*www.realitysand wich.com*) and the social network Evolver.Net (*www.evolver.net*). Daniel Pinchbeck, author of the bestselling book *2012: The Return of Quetzalcoatl*, is one of the founders of both of these sites. His current focus is on promoting integrated political and social thinking to help facilitate the rapid growth of an effective network of people committed to taking on the challenges of preparing for 2012.

Getting further educated about the shifting paradigm of 2012 is becoming progressively easier with a number of groundbreaking independent documentary films that have been made on the subject. These include:

- *2012: The Odyssey* and *Timewave 2013*: Two pioneering documentaries by Sharron Rose and Jay Weidner, author of *The*

Cross of Hendaye, that feature interviews with 2012 theorists from around the world (*www.2012theodyssey.com*).

- *2012: Science or Superstition*: A documentary by Disinformation films that features John Major Jenkins, Daniel Pinchbeck, and Walter Cruttenden (*www.2012dvd.com*).
- *Time of the Sixth Sun*: Dividing its subject matter into seven chapters for each of the chakras, *Time of the Sixth Sun* combines interviews with scientists and indigenous elders about 2012 (*www.timeofthesixthsun.com*).
- *2012: Time for Change:* A feature-length documentary by Joao Amorim that combines film and animation and features Pinchbeck interviewing scientists, anthropologists, physicists, and celebrities about 2012 (*www.2012timeforchange.com*).

101. In a Word: Planetization

If one single word could describe all of the diverse theories competing to define the 2012 mindshift, it would probably be another concept first proposed by Chardin: planetization. The idea of planetization places humanity and the biosphere into a new broader context of a symbiotic unity. To become planetized is to adopt new ways of thinking that reflect a holistic viewpoint between humankind and the planet.

A planetized view is one able to conceptually grasp that this planet is just one of many, from one solar system, among billions of star systems, among billions of galaxies. The Mayan calendar and the Vedic Yuga cycle, with their galactic frames of reference, are both planetized calendars.

The New Copernican Revolution of planetization shifts the center point of our universe from the sun and our local solar system to the new reference point of the galactic center. This dramatically different sense of scale encourages more holistic, global-scale thinking. A rapid shift to a planetized perspective may in fact be the best way to save our planet and ourselves from rampant technological consumerism.

Planetization could be described as almost the polar opposite of globalization. The idea of globalization was to extend a free market around the world to increase international trade. In practice, however, this has tended to favor large multinational corporations at the expense of local, more sustainable industries. The economic transformation of globalization was only sustainable in a period of unprecedented global economic growth. The dual forces of the current economic crisis and ongoing energy crunch mean that the system of manufacturing parts of a car in China, shipping them to Europe to be assembled, and then sending them to Australia or Argentina to be sold no longer makes sense.

If globalization was the de facto economic religion of the expansionist late twentieth and early twenty-first centuries, planetization is the balancing force of the emerging holistic paradigm of 2012. Dramatic choices await humanity as we approach 2012. As the Chinese proverb says, "We are born in interesting times." It will be up to us collectively to decide whether this is a curse or a blessing.

APPENDIX A

Glossary

hau
Yucatec Mayan name of the twentieth Tzolkin day sign, the Sun.

Ajq'ij
Quiché Mayan name for a calendar day keeper.

Bacab
The gods of the four directions.

Baktun
A calendar period of just under 400 years.

Big bang
The idea that the universe has explosively expanded from a condensed state at some time in the past and continues to expand to this day.

Big crunch
A possible scenario for the ultimate fate of the universe, in which the expansion of space eventually reverses and the universe collapses, ultimately ending as a black-hole singularity.

Biosphere
The global sum of all ecosystems. It can also be called the zone of life on Earth.

Black dwarf
A hypothetical stellar remnant, created when a white dwarf becomes sufficiently cool and no longer emits significant heat or light.

Black hole
A region of space where the gravitational field is so powerful that nothing, including light, can escape its pull.

Cahib xalcat be
The four junction roads of Mayan mythology.

Carrington event
The biggest solar flare in the 160-year recorded history of geomagnetic storms.

Catabolic collapse
John Michael Greer's definition of what happens to a civilization when it can no longer meet the demands for resources and energy to support itself. It then begins the process of feeding on its own parts.

Chandler wobble
A small motion in Earth's axis of rotation, relative to Earth's surface, which occurs because Earth is not a perfect sphere.

Chaos theory

The behavior of certain dynamic systems whose states evolve with time. Usually highly sensitive to initial conditions (popularly referred to as the butterfly effect).

Chilam Balam

The order of jaguar priests responsible for prophecy.

Chol'qij

Quiché Mayan word for the 260-day Tzolkin count.

Colony collapse disorder (CCD)

A phenomenon in which worker bees from a beehive or European honeybee colony abruptly disappear.

Cosmic rays

Energetic particles originating from outer space that impact Earth's atmosphere. Almost 90 percent are protons, 9 percent are helium nuclei, and about 1 percent are electrons.

Crop circles

Also known as crop formations; patterns created by the flattening of cereal crops such as wheat, barley, rapeseed, rye, and corn. These often take the form of intricate geometric patterns.

DNA
Deoxyribonucleic acid, a nucleic acid that contains the genetic
instructions used in the development and functioning of all
known living organisms.

Ecliptic
An imaginary circle projected into the sky from the plane of the
solar system.

Eschaton
The end of everything, especially time; the final destiny of the world.

Equinox
An equinox occurs when the sun is located vertically over the
equator. This happens twice a year. The first one, on March 20 or
21, is the spring or vernal equinox for the northern hemisphere
and the autumn equinox for the southern hemisphere. The second
equinox takes place on September 22 or 23

Galactic alignment
The alignment of the winter solstice sunrise with the Galactic equator.
This alignment occurs as a result of the precession of the equinoxes.

Gamma rays
High-energy electromagnetic radiation that is produced by
subatomic particle interactions, such as radioactive decay.

Geosphere

The densest parts of Earth's strata, mostly consisting of rock and regolith.

Gothenburg magnetic flip

A 180-degree flip in the geomagnetic pole of Earth that happened between 12,000 and 13,000 years ago.

Haab

The yearly 365-day calendar.

Hablatun

The largest named period of the calendar. A period of 1.26 billion years.

Heliocentric

Having the sun as the center.

Heliosphere

An elongated bubble in space blown into the interstellar medium by the solar wind.

Hunab Ku

"One god," or "the giver of movement and measure."

Hun Yecil
The flood that destroyed the last world.

Ionosphere
The uppermost part of the atmosphere, distinguished because it is ionized by solar radiation.

Itzá
The postclassical-period invaders of the Yucatán peninsula.

Izapa
Pre-Mayan site in the south of Mexico considered the possible birthplace of the Long Count.

Kepler's Third Law
The square of the orbital period of a planet is directly proportional to the cube of the semi-major axis of its orbit.

Kin
A day; also used as a mantra.

Kuiper belt
A region of the solar system extending from the orbit of Neptune to twice that distance. Similar to the asteroid belt, although twenty times larger.

Local interstellar space medium (LISM)
The relative density of the gas and dust that pervades interstellar space.

Long-period comet
Comets whose orbits are longer than 200 years.

Lunisolar
Relating to the relationship between the moon and the sun.

Magnetosphere
A highly magnetized region around Earth that protects the planet from incoming cosmic radiation.

Maunder minimum
The name given to the period roughly from 1645 to 1715, when sunspots became exceedingly rare and temperatures were unusually low. Also known as the Little Ice Age.

Mayapan
Important postclassical city in the Yucatán.

Meridian
A meridian is one of the vertical lines used to circumscribe the world, such as the prime meridian (Greenwich meridian).

Moore's Law
A description of the long-term trend in the history of computing hardware toward exponential growth in cost performance.

Neutron stars
A remnant that can result from the gravitational collapse of a massive star during a supernova event. Such stars are composed almost entirely of neutrons.

Node
The joint on a stem where a leaf is attached.

Noosphere
A mental envelope or thinking layer that surrounds the atmosphere of the planet containing the totality of all thought forms.

Novelty
The quality of being new. Novelty theory claims this quality can be objectively measured.

Oort cloud
A spherical cloud of comets nearly a light year from the sun.

Physical pole shift

A situation in which the planet actually rolls over on its axis. A physical pole shift would likely be catastrophic for the global ecology. One probable consequence would be major crustal displacement, as the flip causes tectonic plates and continents to collide with each other.

Plasma

Partially ionized gas in which a certain proportion of electrons are free rather than being bound to an atom or molecule. Responds strongly to electromagnetic fields.

Plasmoids

A coherent structure of plasma and magnetic fields. Occur in natural phenomena such as ball lightning.

Precession of the equinoxes

A gradual shift in the orientation of Earth's axis of rotation that traces out a conical shape in a cycle of approximately 25,771 years.

Quetzalcoatl

The Aztec name for the feathered serpent.

Quiché

Strongly traditional Mayan tribe living in Guatemala.

Seyfert galaxies
A subclass of galaxies with active galactic nuclei that appear to be in the process of exploding.

Solar maximum
Time period of maximum sunspot counts.

Sunspot cycles
Sunspot activity follows a cycle that quickly rises and more slowly falls over about eleven years. Significant variations of the eleven-year period are known over longer spans of time.

Sunspots
Magnetic storms on the face of the Sun that form areas of reduced surface temperature.

Transient luminous event
A short-lived, fluorescent, electrical phenomenon that occurs above storm clouds; less commonly called upper-atmospheric lightning.

T Tauri stars
A class of variable stars found near interstellar dust clouds.

Tun
A calendar period of just 360 days.

Tzolkin
The Yucatec Mayan word for the 260-day count.

Uayeb
The five days of purification at the end of the yearly Haab calendar; considered unlucky.

Uinal
A twenty-day period of the Haab or Tun calendar.

Van Allen radiation belts
Bands of plasma around Earth held in place by Earth's magnetic field. There are two main belts: the inner one is mostly composed of protons; the outer one is composed mostly of electrons.

Waveform
The shape and form of a graph plotting certain points.

White dwarf
A small star composed mostly of electron-degenerate matter. The faint luminosity of a white dwarf comes from the emission of stored heat.

Xibalba
The underworld whose mouth is in the dark rift near the center of the galaxy.

APPENDIX B

Resources

Websites

2012: Dire Gnosis
One of the most comprehensive 2012 websites online
www.diagnosis2012.co.uk

2012 Supplies
Survivalist supplies store and discussion forum with lots of
2012 stories
www.2012supplies.com

Alignment2012
John Major Jenkins's website
www.alignment2012.com

Authentic Maya
Lots of information about the history and culture of the Mayan
people
www.authenticmaya.com

Binary Research Institute
Home website for Walter Cruttenden's Conference on Precession
and Ancient Knowledge
www.binaryresearchinstitute.org

BLT Research
Website of the BLT crop-circle research team
www.bltresearch.com

Crop Circle Connector
The world's foremost crop circle website with news and pictures of all the latest formations
www.cropcircleconnector.com

Crop Circle Science
The website of crop circle artist and researcher Andreas Mueller
www.cropcirclescience.org

Daily Galaxy
A great source of space discovery stories and cutting-edge astronomy
www.dailygalaxy.com

Divine Cosmos
David Wilcock's site, an excellent web resource with many in-depth articles on the emerging science of 2012
www.divinecosmos.com

Earth Changes
Michael Mandeville's site tracking the incidence of earthquakes and other earth changes
www.earthchanges-bulletin.com

Enterprise Mission
Richard Hoagland's site, which includes hyperdimensional and torsion physics information relevant to 2012
www.enterprisemission.com

Evolver
A social network site for conscious collaboration that also hosts offline 2012 events
www.evolver.net

Foundation for the Law of Time
Jose Argüellés's official website
www.lawoftime.org

Global Coherence Initiative
A science-based initiative to shift global consciousness from instability and discord to balance and cooperation
www.glcoherence.org

Global Consciousness Project
Multidisciplinary collaboration at Princeton University to study the effects of consciousness on the planet
http://noosphere.princeton.edu

Institute of HeartMath
Research Institute dedicated to studying the effects of human coherence and heart-based living
www.heartmath.org

Maharishi University of Management
Research into the Maharishi effect of Transcendental Meditation
www.mum.edu/m_effect/

Mayan Majix
Information on the Carl Calleman interpretation of the Mayan Calendar on a site founded by Ian Xel Lungold
www.mayanmajix.com

Peter Russell
Lots of interesting visualization tools relating to Russell's global brain theory including a life expectancy calculator and world clock
www.peterrussell.com

Planet Art Network
Global web portal for the Thirteen-Moon Calendar Change Movement
www.tortuga.com

Project Condign
Download site for the declassified UK Ministry of Defence
UFO reports
www.mod.uk/DefenceInternet/FreedomOfInformation/
PublicationScheme/SearchPublicationScheme/
UnidentifiedAerialPhenomenauapInTheUkAirDefenceRegion.htm

Reality Sandwich
Web magazine and multichannel blog site with lots of 2012
content
www.realitysandwich.com

The Sirius Research Group
Binary star theorists who believe our Sun is twinned with Sirius
www.siriusresearchgroup.com

Space Weather
News and information about the current Sun–Earth environment,
including sunspot cycles, near-miss asteroids, and more
www.spaceweather.com

Sphinx Stargate
Website of Paul LaViolette, which includes lots of information on
a broad range of subjects from archaeoastronomy to subquantum
kinetics
www.etheric.com

S.P.I.R.I.T.
Scientific Paranormal Investigative Research Information and Technology is David M. Rountree's organization dedicated to the collection, documentation, and analysis of paranormal phenomena
www.spinvestigations.org

Thirteen-Moon Calendar
Natural time calendars based on *Dreamspell*
www.13moon.com

Time Surfer
Excellent free program for following the Traditional, Long Count, and *Dreamspell* versions of the Mayan calendar for Mac OS X
www.gaianmysteryschool.com/timesurfer/

Tribe.Net 2012 discussion group
One of the best and longest-running discussion boards about 2012
www.2012.tribe.net

Recommended Reading

Argüellés, Jose and Stephanie South. *Cosmic History Chronicles Vol. 1–4*. (Ashland, OR: Law of Time Press, 2004–2008)

A seven-part series, released one each year until 2012, giving an in-depth view of Argüellés's ideas.

Argüellés, Jose. *Earth Ascending*. (Boulder, CO: Shambhala, 1984)

A treatise of the whole Earth that synthesizes the Tzolkin, *I Ching*, and the philosophies of Teilhard and Reiser.

Argüellés, Jose. *The Mayan Factor*. (Santa Fe, NM: Bear & Co., 1987)

The book that brought the Mayan calendar to popular awareness.

Argüellés, Jose. *Time and the Technosphere*. (Rochester, VT: Bear & Co., 2002)

A treatise on the ecological crisis caused by global industrial society and the need to change the world's calendar.

Calder, Nigel and Henrik Svensmark. *The Chilling Stars*. (Cambridge, MA: Icon Books, 2007)

A guide to the idea that cosmic radiation plays a major part in climate change on Earth.

Calleman, Carl. *The Mayan Calendar and the Transformation of Consciousness.* (Rochester, VT: Bear & Co., 2004)

> Calleman's theories about the parallels between Mayan pyramids and their calendar.

Craine, Eugene R. and Reginald C. Reindorp. *Codex Perez and the Book Of Chilam Balam of Mani.* (Oklahoma: University of Oklahoma, 1979)

> A compilation of Mayan prophecies and writings from after the European invasion.

Cruttenden, Walter. *Lost Star of Myth and Time.* (Los Angeles: St. Lynn's Press, 2005)

> A complete guide to the binary star theory and its relationship to Yukteswar's interpretation of the Vedic Yuga cycle.

de Chardin, Teilhard. *The Phenomenon of Man.* (New York: Harper Perennial, 1975)

> The most important exposition of the philosophy of Teilhard de Chardin.

de Santillana, Giorgio and Hertha von Dechend. *Hamlet's Mill.* (Boston: David R. Godine, 1992)

> A compilation of myths about precession from around the world.

Duncan, David Ewing. *The Calendar*. (London: Fourth Estate, 1998)

The story of the development of the Gregorian calendar and the modern concept of time.

Geryl, Patrick. *How to Survive 2012*. (Kempton, IL: Adventures Unlimited Press, 2007)

A theory that a global cataclysm is imminent in 2012 and what to do in order to survive it.

Greer, John Michael. *The Long Descent*. (Gabriola Island, B.C.: New Society Publishers, 2008)

A study of the overshoot and collapse of previous civilizations in the light of the current peak oil energy crisis.

Jenkins, John Major. *Galactic Alignment*. (Rochester, VT: Bear & Co., 2002)

Jenkins explores the galactic alignment in other cultures and philosophical traditions.

Jenkins, John Major. *Maya Cosmogenesis 2012*. (Rochester, VT: Bear & Co., 1998)

The definitive guide to the theory of galactic alignment.

Jenkins, John Major. *The Story of 2012.* (New York: Tarcher Penguin 2009)

> Jenkins chronicles the history and development of the 2012 movement in his new book.

Johnson, Kenneth. *Jaguar Wisdom.* (St. Paul: Llewellyn Publications, 1997)

> A guide to working with the traditional Mayan calendar with much information about Mayan folklore.

Joseph, Lawrence E. *Apocalypse 2012: A Scientific Investigation into Civilization's End.* (New York: Morgan Road Books, 2008)

> A personal journey through theories about 2012 that includes a rare interview with Alexey Dmitriev.

Kurzweil, Ray. *The Singularity Is Near.* (London: Viking, 2005)

> A futurist's view of what life might be like as we approach the technological singularity.

Laszlo, Ervin. *The Chaos Point.* (London: Piatkus Books, 2006)

> An analysis of global society at the tipping point of 2012.

LaViolette, Paul. *Earth Under Fire*. (Portland: Starlane Publications, 2000)

> The galactic superwave theory and its relationship to cycles of catastrophe on Earth.

Lovelock, James. *Gaia: A New Look at Life on Earth*. (Oxford: Oxford University Press, 1982)

> Lovelock's guide to the hypothesis that Earth should be considered as a single living system.

Makemson, Maud Worcester. *The Book of the Jaguar Priest: A Translation of the Book of Chilam Balam of Tizimin*. (New York: Henry Schuman, 1951)

> A translation of an important book of Mayan prophesies.

McKenna, Terence. *The Invisible Landscape: Mind, Hallucinogens, and the* I Ching. (San Francisco: Harper San Francisco, 1994)

> The story of the timewave-zero hypothesis.

McKenna, Terence. *True Hallucinations*. (San Francisco: HarperOne, 1994)

> The travelogue of the McKenna brothers as they journey into the Amazon in search of hallucinogenic plants.

Pinchbeck, Daniel. *2012: The Return Of Quetzalcoatl.* (New York: Jeremy Tarcher/Penguin, 2006)

A personal travelogue through psychedelic awakening into a journey of synchronicity and discovery of 2012.

Pretchel, Martin. *Secrets of the Talking Jaguar.* (New York: Tarcher, 1999)

An account of life as an initiated Mayan day keeper in a traditional Mayan community.

Pringle, Lucy. *Crop Circles: The Greatest Mystery of Modern Times.* (New York: Thorsons Publishers, 2000)

A good introduction to the phenomenon of crop circles and the mystery that surrounds them.

Reiser, Oliver L. *Cosmic Humanism.* (Rochester, NY: Schenkman Books, 1966)

An overview of Reiser's work, including the psi bank and concepts of the role of magnetism in evolution.

Rennison, Susan Joy. *Tuning the Diamonds.* (Burton-upon-Trent: Joyfire Publishing, 2006)

An in-depth look at the relationship between emerging science and subtle energies.

Russell, Peter. *The Global Brain Awakens*. (Shaftesbury: Element Books, 2000)

The theory that humanity's true purpose is to create a telepathic brain for planet Earth.

Shearer, Tony. *Beneath the Moon and Under the Sun*. (Albuquerque, NM: Sun Publishing Company, 1975)

The original source for the harmonic convergence and the first modern reinterpretation of the Tzolkin calendar.

Spilsbury, Ariel and Michael Bryner. *The Mayan Oracle: Return Path to the Stars*. (Rochester, VT: Bear & Co., 1992)

A card set and accompanying book that gives many correspondences for working with the day signs of the Tzolkin.

Stray, Geoff. *2012 in Your Pocket*. (Virginia Beach: A.R.E. Press USA, 2009)

A summary in miniature of Stray's *Beyond 2012*.

Stray, Geoff. *Beyond 2012: Catastrophe or Awakening*. (Rochester, VT: Bear & Co., 2009)

A comprehensive compendium of theories and ideas regarding 2012, as well as the conclusions of the author.

Stray, Geoff. *The Mayan and Other Ancient Calendars.* (Glastonbury: Wooden Books, 2007)

> A simple introduction to how the Mayan calendar works and its relationship to other ancient calendars.

Tedlock, Barbara. *Time and the Highland Maya.* (Santa Fe: University of New Mexico Press, 1992)

> An account of life with the traditional indigenous Maya.

Tedlock, Dennis. *Popol Vuh: The Mayan Book of the Dawn of Life.* (New York: Simon & Schuster, 1996)

> A translation of the Quiché Mayan classic of Mayan mythology.

Thompson, J. Eric S. *The Rise and Fall of Maya Civilization.* (Norman: University of Oklahoma Press, 1954)

> A good introduction to the history and culture of the Mayan people.

Various authors. *The Mystery of 2012: Predictions, Prophecies and Possibilities.* (Louisville, CO: Sounds True Inc., 2008)

> A series of essays from a number of contemporary authors about 2012.

Vernadsky, Vladimir. *The Biosphere*. (New York: Springer, 1998)

Vernadsky's ideas about the biosphere and noosphere.

Yaxkin, Aluna Joy. *Mayan-Pleiadian Cosmology*. (Mt. Shasta, CA: Hauk'in, 1995)

An intuitive guide to working with the energies of the twenty day signs of the Tzolkin.

Yukteswar, Sri Swami. *The Holy Science*. (Self-Realization Fellowship, 1894)

The original source of Yukteswar's reinterpretation of the Vedic Yuga cycle.

About the Author

Mark Heley has been a pioneering researcher of Mayan culture and the theories surrounding 2012 for nearly a decade. He is the producer and director of the 2012 documentary called *Frequency Shift* and has spoken at events and conferences across the United Kingdom, Canada, Europe, and the United States on the subject. He has been a professional journalist for twenty years and has an honors degree in philosophy from Cambridge University. He lives in Glastonbury, England.

BEYOND HERE

Sure, this world is fascinating, but
what's beyond is even more intriguing...

Want a place to share stories and experiences about all things strange and unusual? From UFOs and apparitions to dream interpretation, the Tarot, astrology, and more, the **BEYOND HERE** blog is the newest hot spot for paranormal activity!

Sign up for our newsletter at
www.adamsmedia.com/blog/paranormal
and download our free Haunted U.S. Hot Spots Map!